THE LIBRARY
ST. MARY'S COLLEGE OF MARYLAND
ST. MARY'S CITY, MARYLAND 20686

D1608885

MINERAL RESOURCES
Geology, Exploration, and Development

William H. Dennen

Taylor & Francis
New York • Philadelphia • Washington D.C. • London

USA	Publishing Office:	Taylor & Francis • New York 79 Madison Avenue, New York, NY 10016
	Sales Office:	Taylor & Francis • Philadelphia 242 Cherry St., Philadelphia, PA 19106-1906
UK		Taylor & Francis Ltd. 4 John St., London WC1N 2ET

Mineral Resources: Geology, Exploration, and Development

Copyright © 1989 Taylor & Francis • New York

All rights reserved. No part of this publication may be reproduced, stored in a retrieval system, or transmitted, in any form or by any means, electronic, electrostatic, magnetic tape, mechanical, photocopying, recording or otherwise, without the prior permission of the copyright owner.

First published 1989
Printed in the United States of America

Library of Congress Cataloging in Publication Data

Dennen, William H.
    Mineral Resources: geology, exploration, and development
William H. Dennen.
    p. cm.
    Includes index.
    ISBN 0-8448-1569-1.
    1. Mines and mineral resources.   I. Title.
TN145.D46 1988
622--dc19   88-21672
                                                                 CIP

*For my father who inspired me,
students who challenged me,
and my wife who supported me*

Contents

Preface ix

Part I Geologic Considerations

Chapter 1. Minerals and Their Properties 3
               Mineralogy 3
               Physical Properties of Minerals 9
               Formation and Association of Minerals 16

Chapter 2. Geologic Review 22
               Introduction 22
               Magma Crystallization and Igneous Rocks
                   Shapes of Igneous Bodies 28
               Weathering, Erosion, and Sedimentary Rocks 36
               Metamorphism and Metamorphic Rocks 44

Chapter 3. Groundwater and Ore Solutions 49
               Water Underground 49
               Porosity and Permeability 50
               Shallow Groundwater 53
               Deep Groundwater 55
               Ore Solution and Deposition 62
               Hydrothermal Ore Deposits 67
               Deposition of the Ores 74

Chapter 4. Planning, Exploration, and Geologic Evaluation 79
               Introduction 79
               Exploration Geology 80
               Geophysical Exploration Methods 81
               Geochemical Prospecting 87
               Remote Sensing 89
               Drilling 96
               Geologic Evaluation 97

Chapter 5. Classification of Mineral Deposits and Commodities 106
               Scientific Classification 106
               Mineral Commodity Classification 110

Part II Extraction and Milling

Chapter 6. Mining 117
   Introduction 117
   Extraction Methods 118
      General Principles 118
      Rock Blasting 118
      Open Pit Mining and Quarrying 120
      Dredging 123
      Underground Mining 123
      Miscellaneous Extraction Methods 128
         Borehole Extraction 128
         Heap Leaching 131

Chapter 7. Milling 135
   Principles 135
   Processes and Devices 135
      Comminution and Classification 135
      Separation 141
         Separation Principles 144
      Finishing and Transportation 148
   Principles of Mine and Mill Design 150
      Mill Circuits 150
         Head to Tail Design 150
         Tail to Head Design 151
   Waste Disposal and Reclamation 154
   Summary 156

Part III Economics, Regulation, and Trade

Chapter 8. Economic Considerations 161
   Introduction 161
   Supply and Demand 162
   Capital Formation and Cash Flow 165
      The Cost and Time Value of Money 166
      Implications of a Wasting Asset 169
   Operating Organizations 170
   Profitability 170
   Prediction and Risk Assessment 171
   Determination of Lifetime 176
   A Numerical Example 177

Chapter 9. The Legal Framework 182
   Background for Mineral Law 182
   Lands in the Public Domain 183
   U.S. Mining Law 184
   Taxation 187

Contents

Chapter 10. Mineral Resources and Trade 191
    Resources and Reserves 191
    The International Viewpoint 193
    Transportation 198
    Some Examples of Commodities in World Trade 198
        Phosphate Rock 199
        Bauxite 200
        Tungsten 201
        Platinum Group Metals 202
        Zirconia 203
        Mica 203
        Gold 204

Appendices

    Appendix I   *Atomic Parameters* 211
    Appendix II *Mineral Determinative Tables* 214
    Appendix III *Market Specifications* 238
    Appendix IV *Units and Conversions* 244

Indexes 247
About the Author 255

Preface

Minerals free for the taking have provided human beings with tools, decoration, and raw materials from earliest times and have provided the fundamental underpinning to our increasingly complex technology. The development of the human race and the growth of civilization are in large part the story of the utilization of natural resources, whereas national policy and war have often been dictated by their distribution. Some resources such as foodstuffs, fiber, timber, and water renew themselves at rates comparable to fractions of human lifetimes, but mineral resources and fossil fuels require geologic time for their formation and are effectively nonrenewable. It is this aspect of mineral resources coupled with their erratic distribution that demands that mineral exploitation be carried forward in a careful and conservative manner with due regard to the possibilities of recycling and substitution.

The impact of recent energy shortages—actually failures in planning, production, and distribution rather than lack of raw materials—has been particularly evident because fuels are in direct and immediate use by individuals. The critical short supply of many other mineral commodities, however, is not generally recognized because their role in technology is unseen by the average person. There is today evident need to maintain a satisfactory discovery and development rate for mineral resources because no amount of conservation, substitution, or recycling can completely offset the continuing and increasing need for these nonrenewable materials. It should be remarked that the minerals industry of the United States involves a surprisingly small work force and capitalization compared with manufacturing, service, or government, yet is absolutely vital to the whole economy.

Mineral resources may be defined as deposits of those nonrenewable inorganic substances that can be won from our planet and usefully employed by people. They may generally be divided into fuel and nonfuel materials, both playing major roles in modern civilization. The means whereby mineral deposits are formed, found, and exploited is the theme of this book; the scope of the text, however, is limited to a consideration of nonfuels, although most of the concepts may be applied to fossil (petroleum, natural gas, oil shale, tar sand, coal, lignite) and nuclear fuels.

The terms "economic geology," "ore deposits," and "mineral deposits" have come to be used interchangeably by geologists to identify the subdiscipline dealing with mineral natural resources. These terms, however, have quite distinct connotations. Mineral deposit study is primarily concerned with the theory of formation of mineral aggregates putatively capable of being exploited with profit; ore deposits with the origin, nature, and distribution of mineral raw materials of commercial interest; and economic geology with ore deposits as financial assets vis-à-vis the economic structure.

This text is concerned with economic geology in the strict sense and more generally with the interplay of the experts—geologists, engineers, lawyers, businesspeople, investors—required to make a financial success of an ore deposit.

To understand the nature and distribution of mineral resources, it is necessary to understand the nature of geologic materials and processes since mineral deposits are but rocks of a special kind formed by ordinary processes and for which recoverable value exceeds recovery costs. Such rocks are commonly termed *ore*. To be used, ore must be extracted from deposits by mining or other recovery methods and usually be beneficiated before its sale or use, thus introducing a need to know the fundamentals of mine and mill design and operation. Finally, since raw materials acquire value when they are separated from the Earth in a usuable or salable form, they are subject to ownership, taxation, statutes, and regulations controlling their exploitation, and the inexorable laws of economics throughout their history from discovery through product recovery, marketing, exhaustion, and reclamation.

An obvious goal of this book is to inform the reader of the nature of ore deposits and the role the geologist plays in their discovery and evaluation. Less apparent is the important need to underscore the intricacies of the mineral industry, the roles played by various specialists, and the kind of problems they must solve. Finally, nongeologists both within and outside the industry should find this cohesive view of the world of mineral deposits of interest.

Part I deals with the nature and properties of geologic materials—minerals and their aggregates as rocks together with pertinent examples—and discusses the planning and execution of geologic exploration and evaluation activities. Part II is dedicated to an examination of extraction and milling principles and design. Part III provides information on the economics of mineral production, regulatory constraints, and world trade. Appendices are included.

Part I

Geologic Considerations

Chapter 1
Minerals and Their Properties

MINERALOGY

The nature and quality of an ore deposit reside ultimately in the kind, amount, and properties of the minerals that comprise it. Except for gases, all of the 90-plus chemical elements may be extracted from one or another of the approximately 3,000 known minerals, and many minerals are themselves materials of significant value. Both chemical composition and physical properties of minerals assume great importance in the evaluation of a deposit and loom large when their impact on the design of extraction and beneficiation procedures needed for their recovery is considered.

The following review of mineralogy is meant only to point out some of the more significant chemical and physical variables that must be taken into account in dealing with mineral materials.

Careful examination of nearly any rock will show it to be made up of one or a few kinds of grains. These are minerals and may be thought of as the basic units of geologic materials, including ores, which in turn are simply special kinds of mineral aggregates—rocks—having commercial value. For some ores the value is in some minor mineral or chemical component and increases with the mineral or component increase; for others the value lies in the purity of the rock and is decreased by contaminating impurities. In either instance, the mineral content coupled with its marketable value determine whether or not the rock is ore.

The constraints of geometry, natural abundance of elements, and levels of energy available in the Earth all combine to limit the number of naturally occurring inorganic atomic aggregates—minerals—to only a few thousand species. Of these, only between 100 and 200 minerals are common and about a dozen truly abundant, that is, are rock formers. It is, however, among the less common minerals that many of economic value are to be found and some attention must be given to them although their variety precludes emphasis.

The fundamental aspects of minerals reside in the kinds of atoms of which they are made, the way the atoms are arranged, and the nature of the bonds that hold them together.

Different kinds of interatomic bonds arise because atoms are composed of electropositive nuclei embedded in electrically leaky envelopes of orbiting and negatively charged electrons. For an electron in an outer orbit, the positive field of the nucleus is more or less effectively blocked by intervening electrons. Outer electrons are thus more or less *shielded* from the field of the nucleus, and when atoms are close together electrons are not only attracted to their own nucleus but to the nucleus of their neighbors. The pull of the nucleus on the outer electrons of an adjacent atom may be so strong that one or more electrons are transferred, leaving the donor atom as a positively charged ion (cation) deficient in electrons, whereas the electron receptor (anion) acquires the excess. Oppositely charged ions are mutually attracted and the bond between them is called an ionic bond. Such bonds arise between ionic pairs whose nuclei are well and poorly shielded by their electronic envelopes.

Most minerals are comprised of ions arranged as alternating electropositive and electronegative components. The positive component is typically one or more kinds of single ions and the negative component may be a single ion or a negatively charged group (radical) involving both + and − ions. Examples are well-shielded sodium and poorly shielded chlorine atoms, which interact to form Na^+ and Cl^- ions and combine in cubic array as halite, NaCl. (See Appendix I for a listing of chemical elements.) Other *ionic structures,* for instance, calcite, $Ca(CO_3)$, are alternations of calcium ions, Ca^{2+}, and carbonate radicals, $(CO_3)^{2-}$, which are in turn comprised of positively charged carbon and negatively charged oxygen ions.

Electrons in the envelopes of neighboring atoms having equivalent shielding are not permanently transferred, but rapidly oscillate in their allegiance from one to the other atom. Electrons are thus shared between nuclei (following the strict rules of quantum mechanics) and generate a bond called *covalent.* An example is diamond, C, in which each carbon atom shares electrons with four neighboring atoms in a tetrahedral array.

Very poorly shielded nuclei are characteristic of metal atoms that, in aggregate, are capable of mutually removing the outer electrons from their neighbors shells. The result of this interaction is an array of atomic kernels embedded in a mobile electron "gas." Because electrons in metallic solids are free to move under the influence of an external electric field, metals are good electrical (electron) conductors, whereas ionically and covalently bonded substances are nonconductors.

The various kinds of bonding are not mutually exclusive; rather, a combination of types is the rule. Figure 1-1 illustrates this point, and it may be noted that minerals held together by ionic and ionic-covalent bonding dominate. Because of this, mineral classification traditionally groups the individual mineral species into classes based on and named for the electronegative component. Some examples of mineral structures are shown in Figure 1-2 and a scheme of mineral classification is given in Table 1-1.

The kind, arrangement, and bonding of atoms in minerals give the resulting crystalline solid a set of characteristic chemical and physical properties useful in its identification, separation, and use. Different arrangements of the same kind of atoms may have differing properties, as may identical arrangements involving different kinds of atoms. The marked physical differences in compositionally identical graphite and diamond illustrate the important effects of different atomic arrangement and, al-

Minerals and Their Properties

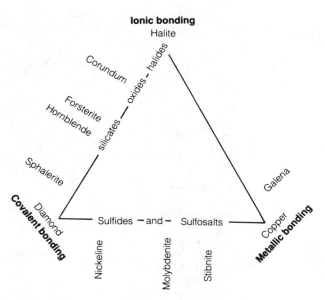

Figure 1.1. The gradational nature of the chemical bond. (*Source:* W. H. Dennen, *Principles of Mineralogy*. New York: Ronald Press, 1960)

though the atoms of hematite, Fe_2O_3, and corundum, Al_2O_3, have an identical arrangement, the properties of these two minerals are quite different because of their different composition.

The more important *physical properties* of minerals are described later and a tabulation of the more common ore and associated minerals is provided in Appendix II. The reader should consider what, if any, advantage these properties lend to the substance in the way of its use or in separating it from other minerals in aggregate.

It must be understood that minerals nearly always show some variation from their theoretical composition and physical properties. Atoms are so small that were they magnified to golf-ball size, a crystal face 1×4 mm would be the size of Tennessee. It is not surprising that perfect arrangement does not exist over such dimensions and that replacement of one kind of atom by another or structural irregularities leading to modified mechanical, optical, or electrical properties takes place. These may be seen, for example, in the presence of minor amounts of nonformulary chemical elements found in assay or in the mosaic of differently oriented patches on the cleavage face of a galena crystal; discontinuities are present between patches but continuity exists from the surface to crystal center.

Solid solution, the interchange of atoms and atoms or vacancies in a mineral structure, is a very common mineralogic phenomenon. The substitution may be a simple replacement of one atom for another, coupled substitutions of atomic pairs in structural sites, stuffing of structural interstices, or substitutions of vacancies for atoms. On occasion, substitution is so widespsread that "solid solution series" is a more appropriate term than is a species designation. Kinds of solid solution are illustrated in Figure 1-3.

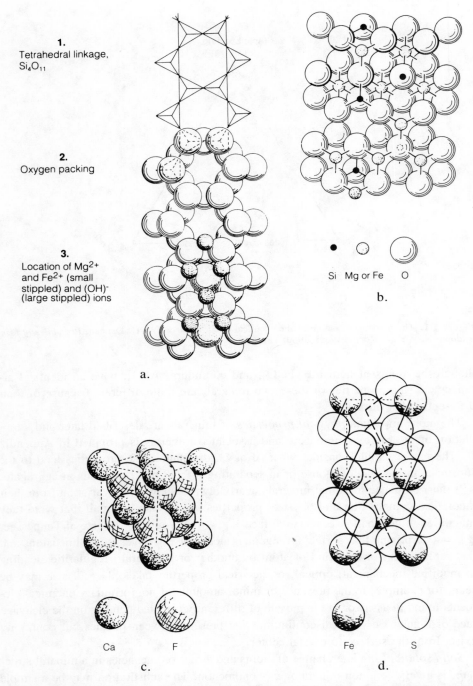

Figure 1.2. Mineral structures. (a) Some structural elements of amphibole, a double chain structure; and (b) Olivine, Mg_2 (SiO_4). (*Source:* W. H. Dennen, *Principles of Mineralogy*. New York: Ronald Press, 1960); (c) Fluorite, CaF_2, (d) Marcasite, FeS_2; (e) Diamond, C.; and (f) Enargite, Cu_3AsS_4. (*Source:* From William H. Blackburn and William H. Dennen, *Principles of Mineralogy*. Copyright © 1988 Wm. C. Brown Publishers, Dubuque, Iowa. All Rights Reserved. Reprinted by permission)

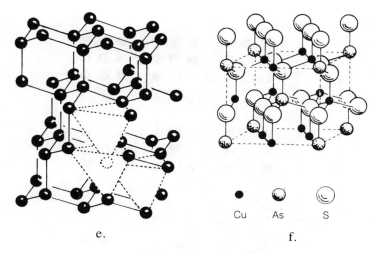

Figure 1.2. *continued.*

Table 1-1.
Scheme of Mineral Classification

Bond Type		Structural Geometry	Mineral Class	Chemical Content
Metallic		Hexagonal close-packed / Cubic close-packed	Native elements	Metallic / Nonmetallic
Covalent		Networks / Sheets / Chains / Molecules	Sulfides / Sulfosalts	
Ionic	No radicals		Halides / Oxides / Hydroxides	Simple / Multiple
	Simple radicals		Carbonates / Sulfates / Phosphates / Vanadates / Tungstates / Molybdates	Anhydrous / Hydrous
	Complex ions	Networks / Sheets / Chains / Pairs / Units / Subsaturates	Silicates	

Source: W. H. Dennen, *Principles of Mineralogy* (New York: Ronald Press, 1960).

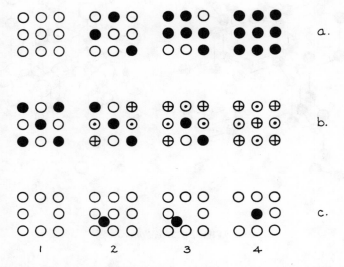

Figure 1.3. Solid solution. (a) Simple substitution. (b) Paired substitution. (c) Miscellaneous: 1. Vacancy for atom, 2. Interstitial atom, 3. Paired interstitial atom and vacancy, 4. Paired atom and vacancy. (*Source:* From William H. Blackburn and William H. Dennen, *Principles of Mineralogy*. Copyright © 1988 Wm. C. Brown Publishers, Dubuque, Iowa. All Rights Reserved. Reprinted by permission)

Atomic substitutions are important in several ways in mineral deposit evaluation and exploration. The substituting element may be valuable even at low concentration: rhenium (xxxx$/lb) may substitute for molybdenum (x$/lb) in molybdenite, MoS_2, thus greatly increasing the value of this mineral; a few percent of chromium makes the difference between clear corundum and ruby; and traces of iron in gibbsite used as aluminum ore could result in excessive refining cost.

Variations in the chemical composition of minerals occur because of the substitution of a chemically similar element for a formulary constituent. In most minerals this substitution is of limited degree, but in some may be complete. Thus the composition of olivine may range from a magnesium to iron silicate with all intermediate compositions; $Mg_2(SiO_4)$ — $(Mg,Fe)_2(SiO_4)$ — $(Fe,Mg)_2(SiO_4)$ — $Fe_2(SiO_4)$. This kind of solid solution can be diagrammatically shown as a bar graph whereon any proportion of magnesium to iron can be represented; 70% magnesium and 30% iron or $Mg_{70}Fe_{30}$ in the example in Figure 1-4.

The nature of solid solution in minerals is often difficult or impossible to assess by the more usual means of observation. Fortunately, it is not the substitutional details that are important but rather the amount of substitution, and this may be found by chemical analysis or assay.

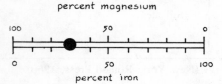

Figure 1.4. Bar graph for the olivine series.

The content of particular elements in an ideal mineral may be readily calculated if the chemical formula and atomic weights of the constituent atoms are known. Each element has a characteristic relative weight with respect to carbon, C, taken as 12.000. (The weights are tabulated in Appendix I.) As an example of the calculations, the copper content of chalcopyrite, $CuFeS_2$, will be:

$$\begin{aligned}
\text{atomic weight of copper, Cu} &= 63.6 \\
\text{atomic weight of iron, Fe} &= 55.85 \\
\text{2 atomic weight of sulfur, S } 2 \times 32.1 &= \underline{64.2} \\
\text{compound or "molecular" weight} &= 183.65 \\
\text{weight \% copper} &= 63.6/183.65 \times 100 = 34.6\%
\end{aligned}$$

If the chalcopyrite content of an ore is 2%, the ore contains $34.6 \times .02 = 0.69\%$ copper metal.

Ideal chemical contents of minerals serve as a first approximation of ore value, but should not replace careful assay for final evaluation. Such assays may be performed by a variety of techniques and represent a field of specialization beyond the scope of this book, although some aspects are discussed in Part II.

PHYSICAL PROPERTIES OF MINERALS

Some of the physical properties of minerals and of their aggregates of importance in their discovery as ores, winning from the Earth by mining, and beneficiation by milling are briefly discussed in this section. Many of the properties used in this context are the same as used in beginning studies; others, however, are little used in mineral identification.

Hardness. In a mineralogic context, hardness is resistance to abrasion. Following Friedrich Mohs (1733–1839), a German mineralologist, it is measured by the use of a series of successively harder minerals arranged on a scale from 1 (softest) to 10 (hardest), each of which will scratch the one below it (Table 1-2). Some other common materials of use in assessing abrasion hardness are also given.

Other measures of hardness, particularly resistance to indentation, are employed by metallurgists. Using such measures, the absolute range of hardness represented by the mineral hardness scale is very large, with diamond being about 1,000 times as hard as talc. Hardness is of particular importance in mineral dressing because of the wear of screens and the jaws and plates of crushers and grinders.

Manner of breaking. Minerals may break by smooth to rough and hackly fractures, deform plastically, or cleave. *Cleavage* is planar rupture controlled by the presence in the mineral grain of planes of lower bond density that arise because of particular atomic arrangements. For example, in Figure 1-5a there are three equally strong bonds per unit length in one direction and four in another, and rupture will

Table 1-2.
Relative Abrasion Hardness

(Mohs scale)

10	Diamond
9	Corundum
8	Topaz
7	unglazed porcelain
7	Quartz
6 1/2	steel file
6	Orthoclase
5 1/2	window glass
5 1/4	knife blade
5	Apatite
4	Fluorite
3	copper
3	Calcite
2 1/2	aluminum
2 1/4	finger nail
2	Gypsum
1	Talc
0.2	wax

preferentially occur in the direction (plane) crossed by the fewest bonds per unit length (area). A similar situation exists in Figure 1-5b because of the unequal spacing of atomic planes. Potential cleavage planes spaced at atomic dimensions are symmetrically disposed within minerals, and different minerals exhibit zero to six directions of cleavage, which result in the different shapes of fragments illustrated in Figure 1-6.

The strength of solids is different for different materials and for different directions of stress application; strongest in compression, intermediate in shear, and weakest in tension. Many devices are in use for the reduction in size or *comminution* of feedstock aggregates, which either place the material under compressison—crush-

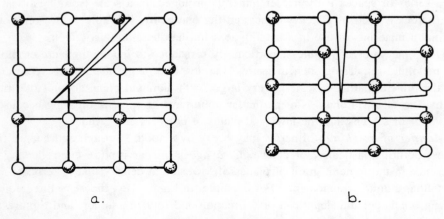

Figure 1.5. Mineral cleavage as a function of bond density (a), and interionic spacing (b).

Minerals and Their Properties

Number of Cleavage Directions	Characteristic Fragment	Example
0		Quartz
1		Muscovite
2		Augite
		Orthoclase
		Hornblende
3		Halite
		Anhydrite
		Calcite
4		Fluorite
6		Sphalerite

Figure 1.6. Shapes of typical cleavage fragments. (*Source:* W. H. Dennen, *Principles of Mineralogy.* New York: Ronald Press, 1960)

ing—or under shear—grinding. Because of the variables involved, some combinations of crushing and grinding stages may be expected to be more effective than others in size reduction, and minerals having different mechanical strength and manner of breaking may be anticipated to break into fragments of different size and shape when crushed together. In aggregates of ore and gangue, this differential breakage may be of prime importance for their separation in mineral dressing. The grains of some aggregates are readily disassociated by crushing and grinding—are free milling—whereas those in other aggregates remain locked (Fig. 1-7a,b). Should the particle size of ore grains and the shape of particles produced by crushing of free-milling ore be as shown in Figure 1-7a, it is apparent that a significant amount of waste could be quickly eliminated by screening (Fig. 1-7c).

The size of particles is particularly important in the mechanical processes used to separate grains of different kinds in mineral dressing. The efficiency of many of the processes used is related to the response of a particle to acclerating forces, gravity or otherwise, so mass (volume × density) and surface area, which establishes resistance to motion through a fluid, are both important.

Two new surfaces are produced each time a fragment is broken so the rate of

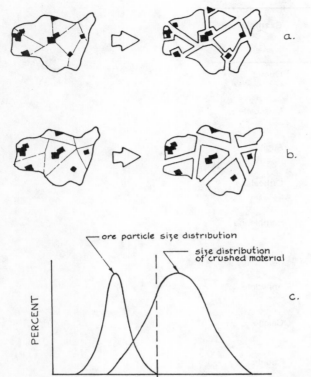

Figure 1.7. Differential breakage. (a) Preferential release of ore minerals in a free-milling ore. (b) Ore minerals remain locked as comminution proceeds. (c) Benefication by screening of a free-milling ore.

production of new area as comminution proceeds is greater than the reduction in volume. Phenomena related to surface area such as solubility and adsorption thus become increasingly effective with finer grinding.

Grain size is usually measured by passage through sieves of known aperture, being recorded as larger (+) or smaller (−) than the mesh. Sets of sieves having apertures in the ratio of the square root of 2 are commonly used in science and industry for the sizing of granular materials. Sieve numbers and apertures for the U.S. Standard Sieve Series are shown in Figure 1-8. The series may theoretically be extended to both coarser and finer sizes; however, in practice there is little need for screening above the coarser openings and practical limitations to sieving below 325 mesh. Size determination of very fine material is usually done by measurement of settling rate in water (elutriation).

Sorting. The effective functioning of many geologic processes and separatory devices used in milling is critically dependent on the uniformity of grain size of the feedstock. If a wide range in particle sizes is present in an ore mineral-gangue aggregate, for example, large lightweight grains and small heavy grains will have the same mass and will travel together rather than be separated as desired. Thus, although a range of particle sizes will always be present in any natural or comminuted

Minerals and Their Properties

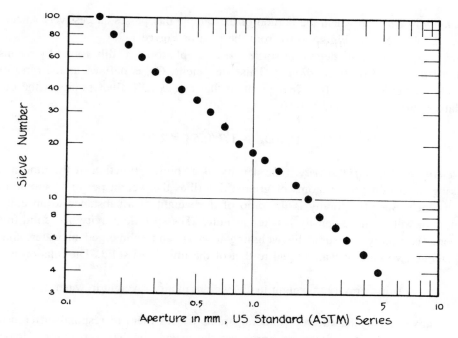

Figure 1.8. Sieve aperature and number.

aggregate, segregation of sizes of materials being milled—*classification*—is an important mineral dressing step.

The determination of the range of sizes present in an aggregate is carried out by placing a weighed sample on the top of a stack of standard sieves with successively smaller apertures, shaking thoroughly, and weighing the fraction of the sample retained on each sieve. The results may be presented as a histogram (Fig. 1-9), in which amounts retained on a particular sieve are identified as (+) and those that pass as (−). In the figure all of the sample is −6 +60 mesh.

A more conventional and useful method of data presentation than a histogram is

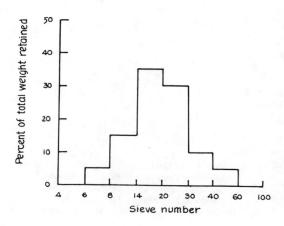

Figure 1.9. Histogram of grain size. (*Source:* From William H. Dennen and Bruce R. Moore, *Geology and Engineering*. Copyright © 1986 Wm. C. Brown Publishers, Dubuque, Iowa. All Rights Reserved. Reprinted by permission.)

a *cumulative curve* in which the portion of the sample that passes successively smaller sieves is plotted as in Figure 1-10 from the data of Figure 1-9.

This curve allows the degree of sorting of a sample to be readily found by means of the *Hazen uniformity coefficient*. This coefficient, C_u, is defined as the ratio of particle diameter at the 60% finer point to that at the 10% finer point on the cumulative curve. From Figure 1-10:

$$C_u = D_{60}/D_{10} = 1.7/0.58 = 2.93$$

Density and specific gravity. The density of a solid, a function of the kind and packing of its constituent atoms or grains, describes the weight per unit volume of a substance. Specific gravity is the ratio of the weight of a substance to an equal volume of water whose density is taken as unity. The specific gravity of a solid free of pores is readily measured by weighing it in air and submerged in water, thus displacing a volume of water equal to that of the immersed solid. The relation is:

Specific gravity = weight in air/weight in air − weight in water

The relative density of different mineral grains causes them to respond differently to identical accelerating forces and provides the means for their separation. Heavy liquid media, either solutions or stirred suspensions of solid particles in a fluid (*suspensoid*), can be used in sink-float separation, and differential acceleration is the mechanical basis for various kinds of separating tables and jigs.

The bulk density or specific gravity of a mineral aggregate or rock is the weighted average specific gravity of the different minerals and void fillings, usually air or water, that are present. This parameter is essential for the calculation of rock or ore tonnage and should be determined by measurement of the material in place. Any method in which the *in-place* weight and volume in a dry or saturated state as appropriate will serve. Common procedures are to remove and weigh the ore from an

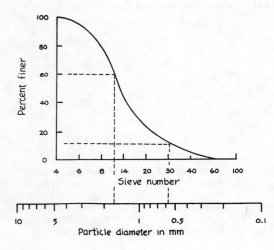

Figure 1.10. Cumulative curve. (*Source:* From William H. Dennen and Bruce R. Moore, *Geology and Engineering*. Copyright © 1986 Wm. C. Brown Publishers, Dubuque, Iowa. All Rights Reserved. Reprinted by permission.)

easily measured channel or length of cylindrical core. Alternately, material from an irregular hole may be weighed and the volume found by backfilling the hole with sand from a calibrated container.

Interaction with light. Minerals may be opaque, translucent, or transparent, may have different surfaces appearances (luster), and may show different colors of bulk and powdered material. *Color* is a complex physical phenomenon that may be a fundamental attribute of a mineral and have its origin in either its structure or bonding, or be incidentally caused by various atomic and structural imperfections. In general, the colors of normally dark-colored minerals will be a constant property, whereas light-colored minerals may exhibit a wide range of generally pale colors. The distinction between fundamental and incidental colors may be made by powdering the mineral, usually by rubbing it on an unglazed porcelain plate, and noting the color of the *streak* of powder.

The distinction of metallic and nonmetallic *luster*—an elusive property of mineral surface appearance related to the refractive index—is easy, but a trained eye is needed to distinguish the glassy, greasy, adamantine, pearly, or resinous qualities of nonmetallic surfaces. Color and luster are particularly important if hand selection (cobbing) is employed for mineral separation as is common for chromite ore, beryl, and large mica books.

Some minerals, for example, diamond, scheelite, fluorite, and some varieties of calcite, will fluoresce when bathed in ultraviolet light, providing another means for their recognition. On occasion light-activated devices may be used to separate them from waste material.

Magnetic and electrical properties. A few minerals, notably magnetite and pyrrhotite, are ferromagnetic, and other iron-bearing minerals may be made so by strong heating, thus providing a means for their separation from nonmagnetic minerals.

Bonding of minerals by ionic, covalent, and metallic mechanisms in different proportions gives them different degrees of electrical conductivity. No free electrons are present in ideal ionic and covalent compounds and they are typically more or less good electrical resistors. Metallic bonds, on the other hand, involve free electrons that migrate in an electric field, so such compounds are conductors.

Nonconductors have the ability to retain an electrostatic surface charge, impossible on a conductor because of its neutralization by electron migration. Separation of the two might thus be accomplished by a device such as is shown in Figure 1-11.

Solubility. Almost all minerals may be dissolved by some one or combination of reagents, but usually the solution times are too long and reagent costs too high to use this method of separation. For some reagent-mineral combinations, however, hydrometallurgical procedures are eminently applicable and this is an active field of investigation. Examples of present applications are the use of weak cyanide solutions to dissolve gold and silver from their ores, oxidizing reagents such as sulfuric acid to liberate copper and uranium, and water to purify halides and other water soluble minerals. Obviously, solution processes work most efficiently on finely divided material wherein all grains of importance have a free reaction surface and are not embedded within a nonreactive grain.

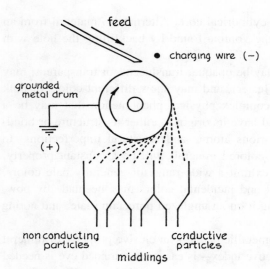

Figure 1.11. Electrostatic separator.

Wettability. The surface of a mineral grain is the locus of distorted atom locations and unsatisfied bonds resulting in physical interaction with liquids to various degrees. Examples are the familiar meniscus formed by liquids in a glass tube, concave downward for water that is attracted by or "wets" glass and convex upward for mercury that does not. This phenomenon is widely exploited in mineral separation through the use of *froth flotation*. Finely divided mineral particles are introduced as a suspension in water or "pulp" to a tank in which bubbles are generated, usually by compressed air and a soapy frothing agent such as pine oil, and nonwetting particles attach themselves to the bubbles. The bubbles with adhering particles rise and the froth is skimmed off, whereas wettable particles sink to the bottom of the tank and are drawn off. Separation may often be improved through the addition of certain organic molecules such as xanthates. These are short-chain molecules whose ends are respectively attracted to oxygen (in the bubble) and metallic surfaces. They thus link metal-bearing minerals to the rising bubbles.

Habit. The external form of a mineral results from the interaction of its internal structure with the environment in which the mineral forms. Habit is not a constant mineral attribute, but particular minerals so often assume a characteristic habit that it is one of the more powerful means of mineral recognition. A glossary of terms used in describing the habits of isolated and aggregated minerals is shown in Figure 1-12.

FORMATION AND ASSOCIATION OF MINERALS

A mineral forms when the atoms in a physical system can cause a net reduction in the system's free energy by combining. It is only required that the essential chemical components are present although the system itself may be closed or open, that is, have a fixed or changing chemical content.

Minerals and Their Properties

Atoms in a gas or liquid undergo random collisions, temporary unions, and dispersion leading to no net aggregation. With falling temperature or increased pressure or concentration, however, aggregation begins. The atoms become fixed in position by interatomic bonds and the regular arrangement of atoms that is assumed is called a crystal structure. (An imaginary coordinate system of points used to describe the positions of matter in a structure is properly termed a lattice; misuse of this term is common.) A clump containing about 200 atoms, about 10 Å in radius, constitutes a stable nucleus upon which further growth occurs.

Crystals may grow by accreting successive layers of atoms, ions, or radicals upon the surface, or by a dendritic mechanism as represented by a snowflake. In this kind of crystallization, spikelike growths extend outward from the nucleus in directions controlled by the crystal structure, side branches protrude at regular intervals, and eventually the interstices are filled. The sequence is analogous to framing and siding a house.

The stability relations of minerals are commonly represented by phase diagrams showing the fields within which particular compounds will form and persist. The more usual parameters relate pressure and temperature (P-T), temperature and composition (T-X), or oxidation and reduction (redox) and pH (see Fig. 1-13a,b,c).

In Figure 1-13a the different forms (polymorphs) of a single component, Al_2SiO_5, are represented. (The components of a system are the minimum number of chemical substances from which it can be prepared, in this instance, one.) The three phases, kyanite, sillimanite, and andalusite, are in equilibrium at the juncture of their fields, the triple point, and any change in pressure or temperature will cause their transformation into a single phase. Because neither pressure nor temperature can change without destroying the equilibrium, there are no degrees of freedom. For a point on any of the field boundaries separating the phases, equilibrium is maintained when either pressure or temperature is changed if accompanied by an appropriate change in the other; there is thus one degree of freedom. Within any of the fields, both temperature and pressure can be changed independently without affecting the equilibrium, and there are two degrees of freedom. In summary, the degree of freedom, F, must equal the number of components, C, plus the physical variables, V, minus the number of phases, P:

$$F = C + V - P$$

This is the phase rule of Gibbs, which also holds in multicomponent systems, and it should be apparent that the maximum number of phases can be present only when the system is invariant, $F = 0$ and $P = C + V$. In many geologic systems there are two degrees of freedom and the number of phases is equal or less than the number of components:

$$P = C + 2 - F; \quad P = C + 2$$

This is the phase rule of Goldschmidt.

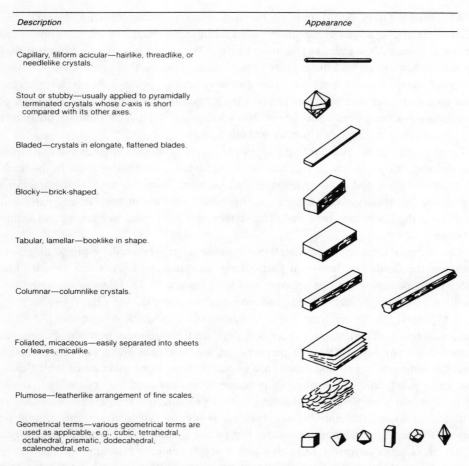

Figure 1.12. Terms used to describe the habits of single crystals, crystal groups, and mineral aggregates. (*Source:* W. H. Dennen, *Principles of Mineralogy*. New York: Ronald Press, 1960.)

The number of minerals in primary association is thus fixed by the chemical complexity (number of components) in the system and the number of effective physical parameters. Molten rock (magma) comprised of eight major components crystallizes into igneous rock with only two to four dominant mineral phases. Sulfide ore solutions must be either more complex chemically, be subject to more physical variables, or be deposited at different times since more phases are commonly associated in the ore-gangue assemblage.

T-X diagrams represent the stability of various chemical compounds as a function temperature, a dominant physical variaible. Figure 1-13b shows the thermal relations of a two-component system and may be understood by following the crystallization history of material represented by point X as temperature falls. Cooling brings the material to point Y on curve D-E. This curve represents the temperature-concentration conditions at which nucleation will occur, nuclei form beneath the curve and dissipate above it. As crystals grow, the composition of the liquid from which they

Minerals and Their Properties

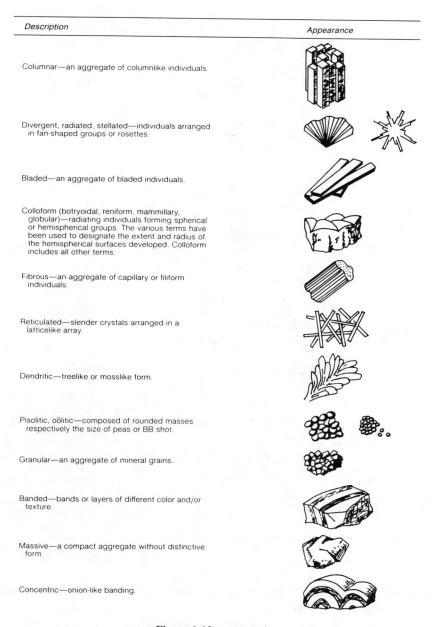

Description	Appearance
Columnar—an aggregate of columnlike individuals.	
Divergent, radiated, stellated—individuals arranged in fan-shaped groups or rosettes.	
Bladed—an aggregate of bladed individuals.	
Colloform (botryoidal, reniform, mammillary, globular)—radiating individuals forming spherical or hemispherical groups. The various terms have been used to designate the extent and radius of the hemispherical surfaces developed. Colloform includes all other terms.	
Fibrous—an aggregate of capillary or filiform individuals.	
Reticulated—slender crystals arranged in a latticelike array.	
Dendritic—treelike or mosslike form.	
Pisolitic, oölitic—composed of rounded masses respectively the size of peas or BB shot.	
Granular—an aggregate of mineral grains.	
Banded—bands or layers of different color and/or texture.	
Massive—a compact aggregate without distinctive form.	
Concentric—onion-like banding.	

Figure 1.12. *continued.*

are being extracted is changed. The crystals are diopside in the example with a composition at the left border of the diagram, and as they form the composition of the liquid becomes more anorthitic, moving to the right along curve D-E. Eventually, the liquid composition will reach point E (the eutectic point) where it lies also on the field boundary for anorthite. At this point the degrees of freedom are reduced

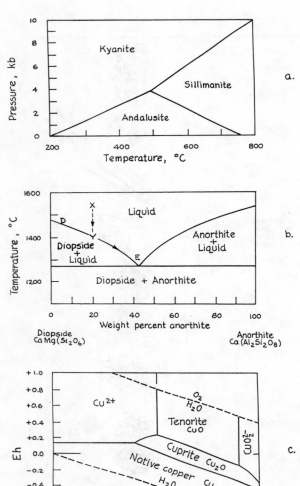

Figure 1.13. Phase diagram. (a) Pressure-temperature relations for $Al_2(SiO_4)O$ minerals. (*Source:* M. J. Holdaway, *American Journal of Science*, 1271: 93–131, 1971) (b) Temperature composition relations for diopside-anorthite. (*Source:* E. F. Osborne, *American Journal of Science*, 240: 751–788, 1942) (c) Eh-pH relations for some copper oxides. (*Source:* R. M. Garrels and C. L. Christ, *Solutions, Minerals, and Equilibria*. New York: Harper and Row, 1965, reprinted by Freeman, Cooper, & Co., San Francisco, CA, 1975)

from two to one, and both diopside and anorthite crystallize at constant temperature until the liquid is exhausted.

Pressure and temperature are dominant physical variables in deep-seated, oxygen-poor geologic systems. Near the surface, however, the ability of atoms to lose or gain electrons—oxidation-reduction or redox—and the activity of hydrogen ions becomes increasingly important. The ease of electron gain or loss is measured relative to hydrogen with the standard potential of the reaction $2H^+ + 2e = H_2$ being taken as zero. Elements with a negative standard potential lose electrons to become positive ions and those having a positive standard potential gain electrons and become negatively charged ions. Because redox reactions take place in aqueous solutions, the effects of the dissociation of water into H^+ and OH^- must also be accounted for; pH, the negative logarithm of hydrogen ion activity, is the common measure.

Figure 1-13c is an Eh-pH diagram showing the stability relations of copper ions and minerals if only copper, water, and oxygen are present in the system. The bounding dashed lines represent the conditions under which water breaks down to oxygen or hydrogen. Note that the phase rule holds and only at a phase boundary may, for example, cuprite and native copper coexist in equilibrium.

The existence of a particular mineral indicates that its crystallization took place within some more or less restricted stability field. Another mineral in primary association with it must have formed under conditions common to both, but since the two fields are unlikely to be identical, the two species could only form together in the overlap portion of their respective stability fields. The more minerals in primary association, the more restricted the conditions of their formation.

Following these general rules, it is clear that mineral associations are not arbitrary groupings of minerals but must be assemblages whose formation, correctly interpreted, leads to an understanding of the conditions of their formation:

Syngenetic deposits are those whose minerals formed at the same time and, therefore, under the same physicochemical conditions.
Paragenetic deposits are comprised of minerals formed successively during a mineralizing event during which the composition, physical parameters, or both underwent continuous change.
Epigenetic deposits have been emplaced in their host rock by its replacement or open-space filling. The epigenetic suite of minerals having been formed at a later time and under different physicochemical conditions from the enclosing rock.

These kinds of deposits are essentially as they were originally formed and are called primary, or *hypogene*. Deposits that have been altered in place to yield a different mineral assemblage are secondary, or *supergene*.

References

Blackburn, W. H., and Dennen, W. H. 1988. *Principles of Mineralogy*. Dubuque: Wm. C. Brown.
Dennen, W. H., and Moore, B. R. 1986. *Geology and Engineering*. Dubuque: Wm. C. Brown.
Deer, W. A., Howie, R. A., and Zussman, J. 1966. *An Introduction to the Rock Forming Minerals*. New York: John Wiley & Sons.
Hurlbut, C. S. Jr., and Klein, C. Jr. 1985. *Dana's Manual of Mineralogy*, 20th ed. New York: John Wiley & Sons.
Mason, B., and Berry, L. G. 1968. *Elements of Mineralogy*. San Francisco: W. H. Freeman.

Chapter 2
Geological Review

INTRODUCTION

In the past few decades geologists have seen the development of the unifying concept of global or plate tectonics. The Earth's crust and upper mantle—the lithosphere—is seen to consist of a few large and a number of small relatively rigid plates (Fig. 2-1). The thermally induced flow of lower mantle material drags on the bottom of the plates and maintains a continuous lateral motion of a few centimeters per year. Plates are added to by material from depth plastered to their trailing edges at mid-ocean ridges and crumpled or consumed at their leading edges. The interaction of the plates results in such features as volcanic and mountain chains, ocean trenches and ridges, and ocean basins.

The dynamic interaction of the plates shapes the continents and oceans and provides the settings in which rocks are formed, destroyed, and reconstituted. Mineral deposits, being but rocks of special character, are also generated and modified; of particular interest here are the thermally induced flow of metal-bearing water underground and the fracturing of rocks to provide channelways for its movement. Much of modern research and writing about mineral deposits has to do with their location and genesis with respect to the various parts of the global tectonic system as it exists today, and particularly in its past configurations.

Figure 2-2 illustrates some of the more important tectonic situations. Subduction carries an oceanic plate to the hot depths where metamorphism followed by selective melting occurs; the lighter liquids rise as intrusive and extrusive magmas with associated thermal and hydrothermal (hot water) effects. The continental plate is arched, broken by faulting, and crowned by a volcanic edifice exposing deep-seated rocks to the water and oxygen-rich surface in a region of considerable relief where weathering and erosion disaggregate, chemically modify, and transport the sediment into the adjacent trench or spread it upon the continental block to be eventually lithified to sedimentary rock.

Analogous actions take place at spreading centers where rising magma differentiates to provide igneous products ranging from chromium-rich ultramafic rocks near the base of the oceanic plate through basalt dikes, flows, and volcanic piles to effusive hydrothermal liquids that deposit their dissolved metals near sea-floor vents

Geologic Review

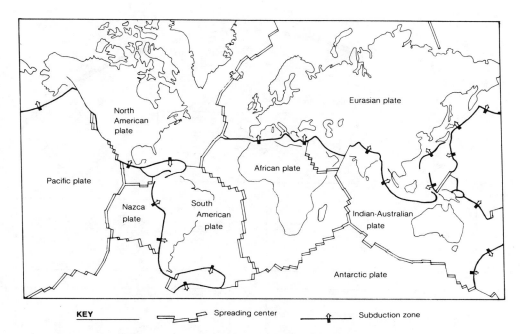

Figure 2.1. Major tectonic plates and their relative motions. (*Source:* Adapted with permission of Macmillan Publishing Company from *The Evolving Earth* by F. J. Sawkins. Copyright © 1978 by Macmillan)

when they come into contact with the cooler ocean water. This differentiated sequence may then be transported by plate movement to be eventually reworked thermally in a subduction zone or be uplifted and eroded in a zone of obduction.

The upper portion of Figure 2-2 is a true scale section through the outer portion of the Earth and the smaller sections show some of the tectonic settings for mineral deposit formation. Section A might be the Pacific margin of South America with a marginal trench and plutons of igneous rock rising under the Andes Mountains. The depth of melting to form the plutons increases away from the ocean margin and the mineral deposits related to their intrusion also changes inland from copper and gold to silver, lead, and zinc to tin, molybdenum, and tungsten.

Section B represents an oceanic spreading center such as the mid-Atlantic ridge. Rising and differentiating plutons can deposit podiform chromite bodies together with nickel and platinum at the base of the lithospheric plate and base metal sulfides are formed when rising solutions meet the overlying ocean waters. Manganese nodules are formed on the thin layer of sediment covering the oceanic lithosphere.

Section C shows the complex relations to be found when an island arc, say Japan, develops outside a back arc basin separating it from the continental mainland. The arc area may contain copper and gold deposits related to deep-seated plutonic rocks and massive base metal sulfide deposits in the volcanic piles that cap the arc. Deposits in the back arc basin are as those in section B and on the continent as in Section A.

A continent-continent collision shown in section D might be represented by the

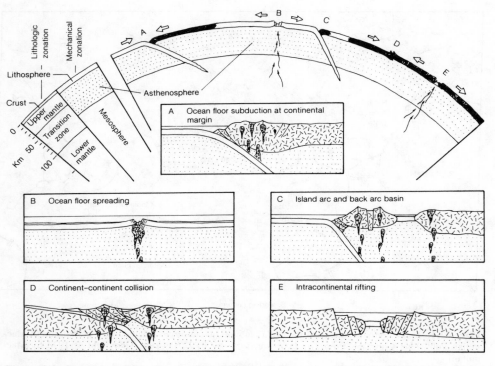

Figure 2.2. Plate interactions. (*Source:* From William H. Dennen and Bruce R. Moore, *Geology and Engineering*. Copyright © 1986 Wm. C. Brown Publishers, Dubuque, Iowa. All Rights reserved. Reprinted by permission)

Himalayan ranges and could contain tin, tungsten, silver, nickel, and cobalt in its magmatic zones, jadeite in the complexly overthrust zone, and uranium in the sediments shed from the upthrust areas.

The Red Sea is representative of an intracontinental rifting, Section E. Metal-rich brines and sediments are to be found within the basin, manganese deposits on submerged steps, and lead-zinc deposits on the subaerial rim.

Ore deposits, as all rocks, have an intimate relationship with their setting in the tectonic regime. Their origin, however, is more readily understood in terms of earth processes. These processes and the materials that comprise the Earth may be epitomized diagrammatically by the "rock cycle" (Fig. 2-3). Here the principal materials of the lithosphere are boxed and the dominant geologic processes that convert them to other products are shown on the arrows. A tabulation of ores and the geologic processes that may be responsible for their concentration into workable deposits is given in Table 2-2.

The Earth-forming materials may have intrinsic value such as use for fill or building stone, but most deposits of value represent special conditions whereby ordinary processes have generated unusual concentrations of minerals. Examples are the concentration in a placer (plăcer) deposit of dispersed gold through the action of running water, or the formation of a residual manganese oxide orebody by removal through weathering of other constituents.

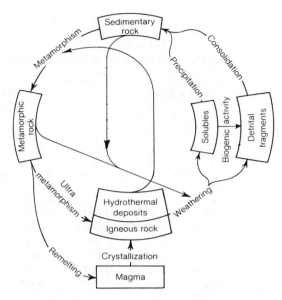

Figure 2.3. The rock cycle. (*Source:* W. H. Dennen, *Principles of Mineralogy*. New York: Ronald Press, 1960)

In effect, the grand processes of the lithosphere may be likened to giant chemical processing plants in which the raw feed is more or less effectively separated into various chemical compounds. The common rock products are seldom if ever chemically pure, but are occasionally sufficiently enriched that they may be economically exploited. Table 2-1a lists the average concentration of a few chemical elements in some of the various rocks that make up the Earth's crust. Table 2-1b shows the average amounts of some elements in the crust as a whole and the natural enrichment needed to generate an orebody. It should be noted that different kinds of rocks contain different levels of elements, so certain rocks are more favored as sources of particular elements.

The importance of understanding both the broad principles and detailed aspects of geologic processes is indicated by Table 2-2. Here a number of metals and nonmetallic substances are correlated with the principal geologic processes that can concentrate them into workable orebodies. Brief descriptions of the processes are given in this and the following chapter. More detailed information is readily available in standard texts on physical geology and mineral deposits.

Rocks are aggregates of minerals and are traditionally classified into three major groupings, each further subdivided as far as needed to distinguish units or bodies for practical or scientific purposes. The subdivision may be made on either a descriptive or genetic basis, the latter implying knowledge of formational history. The major rock groupings are igneous (formed from a melt), sedimentary (deposited, usually in layers or beds on the Earth's subaerial or submarine surface), and metamorphic (transformed by heat and pressure in the solid state from pre-existing rocks). One further division of importance should be included, that of sediments that are unconsolidated soils and rock debris.

Table 2-1.

(a) Average Content of Some Chemical Elements in Different Kinds of Crustal Rocks
(weight percent)

	Igneous Rocks[1]				Sedimentary Rocks[2]		
	ultramafic	mafic	intermediate	salic	shale	sandstone	carbonates
Aluminum	0.5	8.8	8.9	7.7	8.0	3.4	0.8
Iron	9.9	8.6	5.9	2.7	3.1	1.3	0.4
Chromium	0.2	0.02	0.005	0.0025	0.009	0.0035	0.0011
Nickel	0.2	0.016	0.0055	0.0008	0.0068	0.0002	0.002
Manganese	0.15	0.2	0.12	0.06	0.085	0.005	0.11
Copper	0.002	0.01	0.0035	0.002	0.0045	0.0005	0.0004
Lead	0.0001	0.0008	0.0015	0.002	0.002	0.0007	0.0009
Zinc	0.005	0.01	0.01	0.007	0.0009	0.0016	0.002

[1]*Source:* A. P. Vinagradov, *Geokhimiya,* 1962.
[2]*Source:* K. K. Turekian and K. H. Wedepohl, *Bull. Geological Society of America,* vol. 72, 1961.

(b) Average Crustal Composition and Enrichment Necessary to Form an Orebody

Metal	Average Concentration in Crystal Rocks (grams/tonne)[1]	Enrichment Needed to Form Ore
Aluminum	81,300	4
Iron	50,000	5
Titanium	4,400	7
Manganese	950	380
Vanadium	135	160
Chromium	100	3,000
Nickel	75	175
Zinc	70	350
Copper	55	140
Cobalt	25	2,000
Lead	13	2,000
Tin	2	1,000
Uranium	1.8	500
Molybdenum	1.5	1,700
Tungsten	1.5	6,500
Mercury	0.08	26,000
Silver	0.07	1,500
Gold	0.004	2,000

[1]Metric ton, see Appendix III.

Table 2-2
Geologic Processes Effective in Forming Workable Orebodies

	magma crystallization	pegmatite formation	submarine volcanic-exhalative processes	deposition from hydrothermal solutions	weathering	evaporation	sedimentation	placer deposition	metamorphism	contact metasomatism
METALLIC ORES										
Aluminum					O					
Antimony				O						
Bismuth				O						
Chromium	O				O					
Cobalt				O						
Copper	O		O	O			Ø			O
Gold				O				O		O
Iron	O			O	O		Ø			O
Lead			O	O			Ø			O
Lithium		O								
Manganese			O		O		Ø			O
Mercury				O						
Molybdenum				O						O
Nickel	O			O	O					
Platinum	O							O		
Silver			O	O						O
Tin				O				O		O
Titanium	O							O		
Tungsten				O				O		O
Uranium		O		O			Ø			
Zinc			O	O			O			O
Zirconium								O		
NON-METALLIC ORES										
Asbestos									O	
Barite				O	O					
Borates						O				
Clay					O	O				
Corundum										O
Diamonds	O							O		
Feldspar		O								
Fluorspar				O						
Garnet									O	O
Gems		O						O		
Glass sand							O			
Gypsum						O				
Kyanite									O	
Limestone						O	Ø			
Mica		O								
Phosphates	O				O		Ø			

Table 2-2 (*continued*)
Geologic Processes Effective in Forming Workable Orebodies

	magma crystallization	pegmatite formation	submarine volcanic-exhalative processes	deposition from hydrothermal solutions	weathering	evaporation	sedimentation	placer deposition	metamorphism	contact metasomatism
Potash						O				
Salt						O				
Slate									O	
Sulfur			O	O			Ø			
Talc									O	

Ø bacterial involvement possible or essential

Physical and chemical properties, dimensions, shape, orientation, and location are all of great importance when assessing the workability of a potential ore deposit. One far removed from markets must shoulder large transportation costs, and deposits in many areas may not be minable at all because of regulatory controls or political considerations; a steeply dipping bed must be mined in a different and probably more expensive manner than one that lies flat; a body with regular shape allows mining by a cheaper regular plan; and savings of scale may be gained when mining larger bodies.

Mining and milling practice must also take such rock and mineral properties as strength, density, breakability, porosity, and many more into account. The mere identification of a mineral deposit is insufficient to cause great rejoicing by its finder, and an accumulation of minerals is an ore only when it can demonstrably be placed on the market at a profit. The many aspects that must be considered in this regard constitute a recurrent theme discussed at appropriate places throughout the text.

MAGMA CRYSTALLIZATION AND IGNEOUS ROCKS

Magma is subterranean molten rock that is called lava when it appears at the surface. The consolidation of magma on cooling takes place through the interaction of intricate physical and chemical processes occuring over vast scales of space and time, often thousands of cubic kilometers of molten rock cooling over millions of years. Conceptually, this is among the more complex of geologic processes and full appreciation of the system is very difficult, yet understanding is essential for insight into the formation of many mineral deposits.

Magma, generated by the selective melting of previously formed rocks, is a hot, viscous, multicomponent liquid dominated by silicates but containing substantial amounts of water and lesser amounts of low-melting point substances such as carbon dioxide, fluorine, sulfur, and boron. The mutual solution of these many components reduces the freezing point of the liquid mass just as does a mixture of antifreeze and water, so magma is a liquid at temperatures well below the freezing point of its dominant silicates.

Once formed, the density of magma is less than that of solid rock and the liquid mass tends to rise from its birth depth of perhaps 10–50 kilometers to cooler and lower pressure regions in the upper portions of the Earth's crust, or even to be extruded as lava onto the surface.

Cooling of the magma promotes crystallization of mineral grains within the liquid. First one kind of mineral forms and, as temperature falls, it reacts with the remaining magma to generate a new mineral kind. Mineral species thus form and disappear in a more or less orderly reaction series as set forth by Bowen (1928). Two independent reaction sequences are simultaneously present at higher temperatures and converge as temperature falls. In the ferromagnesian (iron and magnesium-bearing silicate minerals) sequence, one mineral after another crystallizes and dissolves with falling temperature in the order: olivine, pyroxene, amphibole, biotite mica. In the plagioclase feldspar* sequence, an early formed aluminosilicate skeleton is retained throughout the crystallization history with particular kinds of ions moving in and out and causing a continuously varying change in composition with falling temperature. The two series converge at a lower temperature (500–600° C) with the crystallization of potash feldspar, muscovite mica, and quartz or, in the absence of sufficient silica, nepheline or a similar silica-poor phase. Thus at any given time the cooling magma contains a fluid phase, stable solids, and dissolving and crystallizing minerals. The continuously varying proportions of the principal mineral constituents are illustrated in Figure 2-4.

The classification of igneous rocks is based on an interpretation of their cooling history as evidenced by their mineral assemblage and fabric (size, shape, and orientation of grains). The reaction series identifies the dominant minerals that should coexist, for example, pyroxene and calcic plagioclase or quartz and sodic plagioclase, and the fabric gives insight to the rate, and hence depth of consolidation.

The reaction series deals only with the dominant silicate minerals, but other components are also present and have similar histories. As crystallization of higher freezing-point constituents proceeds, the residual magma liquid (rest magma) becomes increasingly enriched in such hyperfusible components as water and those constituents, including a number of metals, that prefer, and are thus partitioned to, the liquid phase.

The concept of chemical partition is particularly useful in considering the pathway whereby a particular element becomes concentrated. Matter exists in the solid, liquid, and gaseous states. Should two or more of these states coexist, as is a very common condition, individual atoms (or ions, radicals, molecules) tend to favor residence in one state over another at the particular conditions obtaining. Chemical

---

*Plagioclase feldspars—aluminosilicate minerals with compositions from higher temperature anorthite, $Ca(Al_2Si_2O_8)$, to lower temperature albite, $Na(AlSi_3O_8)$.

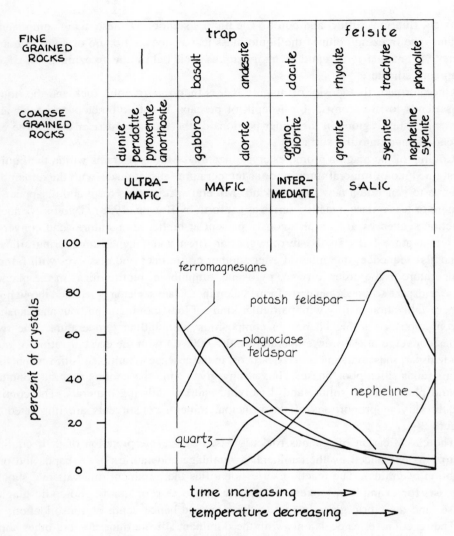

Figure 2.4. Classification of igneous rocks.

elements may thus be preferentially concentrated in the solid, liquid, or gas phase. The ratio of the concentration of an element between two phases is its partition ratio, P.

$$P = C_A/C_B$$

where C_A and C_B are the respective concentrations of a given element in phases A and B. As examples, to say that a substance is water soluble is to say that it will partition selectively to the hydrous phase or, if insoluble, will tend to form or join a solid mineral phase. In cooling magmas, many elements, including a large number of economically important metals, are strongly partitioned to the hydrous phase and so separated from the common rock-forming minerals and enriched in a highly mobile medium.

Early formed crystals may be retained in the final solid rock if the magma is particularly enriched in their constituents either by initial composition or if earlier formed grains are armored by later material enclosing them. Alternatively, heavier crystals may settle through the liquid magma to form lower and upper zones, respectively, enriched and depleted in their constituents (Fig. 2-5). Internal motions in the magmatic liquid may be initiated by the crystallization and sinking of solid phases or differential cooling of the mass. Not only may solids be segregated and removed from reaction by armoring and sinking, but also liquid may escape through porous wall rocks of the magma chamber or react with them to be modified in composition.

The world's greatest concentrations of economic minerals formed by crystal settling in a magma are found in the Bushveld Igneous Complex in the Transvaal of South Africa. The complex covers 67,000 km² and is made up of coalesced basin or funnel-shaped intrusions containing rhythmically banded mafic and ultramafic rocks (Fig. 2-6). This thick layered sequence is divisible into four distinct zones:

Upper Zone	1,200 m thick
Main Zone	3,050 m thick
Critical Zone	1,070 m thick
Basal Zone	1,200 m thick

Within these rocks are tabular cumulate deposits of magnetic iron ore, platinum metals, and chrome ore of remarkable persistence as well as pod, pipe, and dikelike

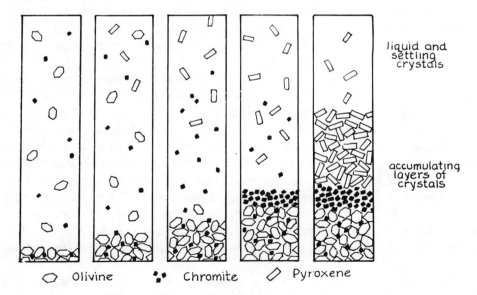

Figure 2.5. Crystal settling in magma. (*Source:* Based on T. N. Irvine and C. H. Smith, "Primary oxide minerals in the Muskox Intrusion," *Economic Geology Monograph* 2, 1949)

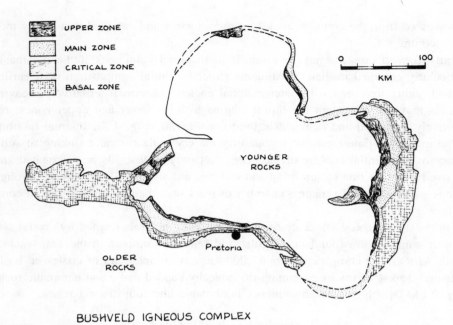

Figure 2.6. Bushveld Igneous Complex. (*Source:* Redrawn after J. Williemse, *Economic Geology Monograph* 4, 1960)

orebodies of the same mineralogy. Twenty-six vanadian magnetite bands occur in the Upper Zone and upper portion of the Main Zone. The Merensky Reef at the base of the Main Zone carries platinum-group metals and sulfides of nickel and copper. This band is only 0.2–6 m thick but can be traced for 250 km on the surface and has been proven by drilling to extend downward for at least 1.8 km. Reserves of at least 10^9 tonnes (metric tons) of chromium ore are present in chromite-rich bands in the basal portion of the Critical Zone.

Continuing consolidation of magma eventually brings the volume of crystals to a point where they are no longer freely floating as individuals or clumps, but are in continuous contact. At this time the magma will no longer respond to mechanical stress as an infinitely deformable liquid, but rather as a brittle solid. Because of its very long cooling time over some millions of years, it is unlikely that any large volume of magma will complete its solidification without being subjected to tectonic stress and, after the stage of embrittlement, such stress may have very important consequences on the course of its fractional crystallization. Cracks forming under moderate stress will be filled with interstitial fluids that crystallize to form cognate dikes, whereas intense stress may cause the expulsion of large volumes of rest magma, which then begins life anew as a liquid mass with a composition quite different from that of the original.

The residual fluids of magmatic fractionation are water-rich, siliceous, and contain those elements that have been partitioned into the liquid, and increasingly hydrous, phase. High water content leads to lowered viscosity and bodies of coarse to very coarse-grained rocks called pegmatites form. They are to be found in regions that have experienced igneous activity such as the Appalachian Mountains (Fig. 2–7). Typically, pegmatites are small, irregularly tabular, lenticular, or branching bodies never more than a few kilometers in length. Those being mined are usually in the range of 2 to 30 by 30 to 300 meters. A rude internal zoning of texture and mineral content is common (Fig. 2-8). Mineralogically, most are dominated by quartz and feldspar together with more or less mica and lesser amounts of uncommon minerals. Feldspar and mica are often exploited and the uncommon constituents include many minerals and elements for which pegmatite may be the only source. A partial list includes the minerals garnet, beryl, tourmaline, and topaz used as gems and minerals containing the rare elements tin, tungsten (wolfram), tantalum, columbium (niobium), uranium, thorium, beryllium, lithium, cesium, and the rare earth elements, particularly cerium, lanthanum, and yttrium. (Four of the rare earth elements—yttrium, ytterbium, terbium, and erbium—take their name from the village of Ytterby, Sweden, where a pegmatite supplied the first samples for study.)

Water has been previously identified as being present in magma, although, because it is a fugitive constituent, little is retained in the cooled and crystallized igneous rock product. The importance of water in magma consolidation and fractionation lies in its fluxing action that increases the mobility of mineral-forming atoms, its ready separation from the magma with lowered pressure, and in the fractionation of certain, often valuable components into any aqueous phase that develops.

Water may be dissolved in magma just as any other component. None, however, is abstracted in minerals in the early stages of consolidation and only small amounts are incorporated in amphiboles and micas at later stages so the magma is progressively enriched in water. The control of its incorporation in silicate melts as a function of temperature and pressure was originally investigated by Goranson (1931) and has been thoroughly documented by a number of later workers. Figure 2-9 shows the amount dissolved in a granitic melt at 900° C as a function of pressure. Since pressure depends upon depth, it may be seen that water will separate from a rising

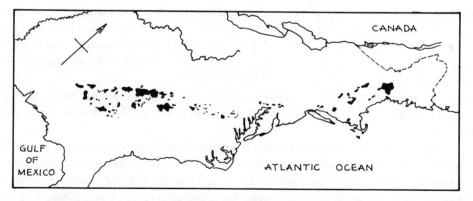

Figure 2.7. Pegmatite deposits in the eastern United States. (*Source:* From E. N. Cameron, R. H. Jahns, A. H. McNair, and L. R. Page, "Internal structure of granitic pegmatites," *Economic Geology Monograph* 2, 1949)

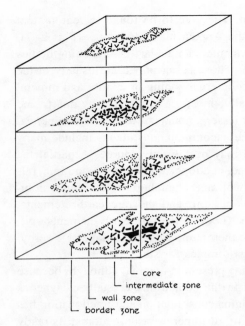

core
intermediate zone
wall zone
border zone

Figure 2.8. Serial sections through an idealized pegmatite body. (*Source:* From E. N. Cameron, R. H. Jahns, A. H. McNair, and L. R. Page, "Internal structure of granitic pegmatites," *Economic Geology Monograph* 2, 1949)

magma in significant quantities. For example, a magma containing 6 weight percent water is undersaturated at a depth of 10 km, point A, and becomes saturated on rising to a depth of 3.8 km, point B. With further rise, the amount of water that can be dissolved in the magma falls and at a depth of 2 km is only 4%. The difference, 2 weight percent, must be exsolved. Having a low viscosity, this water will readily move through the interstices of the crystal mush and penetrate the containing walls. Atoms in the liquid portion of the magma have preference for silicate or aqueous phases and selectively partition into them, thus a migrating aqueous phase carries with it a characteristic suite of dissolved elements that will later be deposited from the hydrothermal solution. Deposits of this kind are to be found throughout the great thickness of rock cover above the magma chamber and represent a particularly important mineralization mechanism. Because of the complexity of the deposits, further discussion and examples are deferred to a later part. However, a sense of scale may be obtained by realizing that if a cubic kilometer of magma exsolves four weight percent of water, this amounts to a volume of about 10^8 cubic meters, enough to cover the District of Columbia knee deep! If only 10 parts per million of a metal is dissolved in this hydrothermal solution, some 1,000 tons would be carried, enough to form an ore vein with a tenor of 5% that was a meter thick, 100 m long, and 80 m deep.

If water is exsolved at relatively high temperature and low pressure such as would exist in magma near the Earth's surface, it will become steam with a very large increase in volume. Bubbles form, expand, and stream upward, carrying crystals and liquid droplets much as a carbonated beverage effervesces when pressure is released on opening its container. This is the driving mechanism for volcanic activity and the results may be observed in the vesiculation of lava, explosive eruption, and

Geologic Review

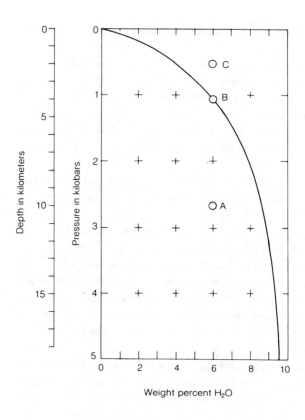

Figure 2.9. Solubility of water in molten granite as a function of pressure at the 900° isotherm. Key: (A) Undersaturated Magma. (B) Saturated Magma. (C) Oversaturated Magma. (*Source:* After R. W. Goranson, *American Journal of Science*, vol. 22, 1931.

abundant steam. Additionally, the presence of a gas provides yet another phase for elemental fractionation and a suite of pneumatolitic minerals may be deposited by sublimation.

Shapes of Igneous Bodies

Magma may undergo consolidation at great or intermediate depths, or on the surface, each level being characterized by different shapes of the rock masses and fabric of the resultant igneous rock.

Deep-seated (plutonic, abyssal) igneous rock bodies are typically very large "bottomless" masses that, when exposed by uplift or erosion, occupy surface areas of hundreds to many thousands of square kilometers. The rock is obviously granular with crystal sizes of 1 to 100 mm and will usually be a granite, granodiorite, or gabbro (Fig. 2-4). These huge bodies are termed batholiths and are usually compound masses comprised of several plutons that have been emplaced nearly contemporaneously. The only other deep-seated body is a pipe, a roughly cylindrical body a few hundreds of meters in diameter but reaching the surface from depths of up to 300 kilometers. Pipes are the conduits for upward-moving magma extruded from volcanic vents and more rarely deeper taps to the mantle. The rock of these latter pipes is always ultramafic, usually peridotite, and may occasionally carry diamonds.

Such pipes are apparently explosively eruptive bodies driven by the exsolution of carbon dioxide gas and steam.

The rock of diamondiferous pipes is called kimberlite; the greatest known concentrations are in South Africa and the Yakutsk Republic in Siberia with smaller clusters on the Colorado-Wyoming border and in northwestern Australia. Occasional isolated pipes are known, for example, in Arkansas, and the extensive alluvial diamond fields such as those found in West Africa, Venezuela, and Brazil suggests their undiscovered presence. The outcrops of diamond pipes are, after all, no more than a few acres in extent and readily eroded.

Shallow or hypabyssal igneous bodies are varied in both shape and size. Most common are tabular masses only tenths to multiple meters thick but often very extensive in the other dimensions. Commonly they are emplaced in real or incipient cracks marking the direction of extension in their host rock and are thus a key to its deformational history. The intrusive rock may be of any composition, but most commonly is basalt or closely related material, fine to medium grain, and often porphyritic. If the tabular mass is emplaced parallel with a pre-existing bedding or foliation of the enclosing rock (is concordant), it is termed a sill. If crosscutting or emplaced in a structureless body, it is a dike. It may be here noted that veins are also tabular masses comprised of one or more minerals but differ from sills and dikes in not having the composition of igneous rocks.

The extrusion of magma onto the surface results in a number of distinguishable masses of igneous rocks. If the eruptive center is a pipe, a conical volcanic pile may be built up by successive eruptions, whereas an elongate volcano or lava plateau forms over feeder dikes and sills. The material extruded may flow out as a liquid sheet or river of lava or be expelled into the air as pyroclastic debris to fall and form more or less regular beds. Surface deposits are thus characteristically tabular or lenticular masses.

WEATHERING, EROSION, AND SEDIMENTARY ROCKS

Rocks, like people, will adjust to new surroundings, and since the differences in temperature and pressure, and amounts of free water, oxygen, and carbon dioxide between the depths in the Earth and its surface are particularly marked, igneous and other rocks are subjected to a radically new environment when they are exposed at the surface by erosion or mechanical movement. The changes that take place are both physical, particularly reduction in particle size, and chemical, involving such reactions as oxidation, hydration, and carbonation. Physical breakdown increases the available surface on which chemical reaction can take place and chemical changes, being expansive, promote disaggregation. Working hand-in-hand, these weathering processes reduce the rock to more or less finely divided debris. Some common minerals of igneous rocks as starting materials and the products formed by their weathering are shown in Figure 2-10. It should be noted that these products may include

Primary minerals and (major elements)	Weathering Products			
	unchanged solids	new solids	colloids	dissolved ions
Quartz (Si)	Quartz (Si)	—	—	—
Feldspar (K, Na, Ca, Al, Si)	—	Clay (Al, Si)	Si, Al	K, Na, Ca
Ferromagnesian minerals (Ca, Mg, Fe, Al, Si)	—	Clay (Al, Si)	Si, Al	Ca, Mg
		Iron oxides (Fe)	Fe	

The weathered product is a mixture in various proportions of iron oxides, clay minerals, and quartz. Soils will also contain more or less organic debris

Figure 2.10. Mineralogical and chemical products of weathering. (*Source:* From William H. Dennen and Bruce R. Moore, *Geology and Engineering*. Copyright © 1986 Wm. C. Brown Publishers, Dubuque, Iowa. All rights reserved. Reprinted by permission, Wm. C. Brown, 1986)

solids (rock fragments and the minerals quartz, clay, and iron oxides), colloidal silica, alumina, and iron, and dissolved ions of calcium (Ca^{2+}), magnesium (Mg^{2+}), sodium (Na^+), and potassium (K^+). Obviously, an effective chemical separation may be performed if solids accumulate as a soil and the soluble ions are leached away in surface and underground water.

Weathering under arctic conditions is mainly physical; temperate climates result in the kinds of products shown in Figure 2-10, and weathering under humid tropical conditions, either past or present, may lead to the dissolution of silicates and limited mobilization of iron and alumina. Tropical weathering may thus lead to the formation of valuable ore deposits on protoliths of appropriate composition. Many of the world's great iron deposits, for example, are cappings of hematite, Fe_2O_3, that have accumulated on an iron and silica-rich sedimentary rock called iron formation from which the silica has been leached. Deposits enriched in metals that have an analogous chemistry to iron such as nickel, chromium, and manganese may also be formed by weathering of the appropriate precursor rocks. Another valuable ore developed by intensive weathering is bauxite, the ore of aluminum, because, under a humid tropical regime, silica may be dissolved from clays leaving an aluminum hydroxide residuum. About two-thirds of the world's bauxite production comes from large deposits of this origin in Australia (31%), the Caribbean (21%), and West Africa (15%). Extensive undeveloped deposits are known in Brazil, and Figure 2-11 illustrates some typical features.

The debris of weathering may be moved away from its point of origin by such agents of transportation as wind, running water, and waves, glaciers, or mass movement. Of these, running water is by far the most important. Particles in moving water may be rolled or slid, be suspended, or saltate (leapfrog). Their manner and thus rate of movement at a given water velocity and turbulence depends critically on the size, shape, and density of the particles, so different grains travel at different rates. Of particular importance to the formation of certain mineral deposits is the marked

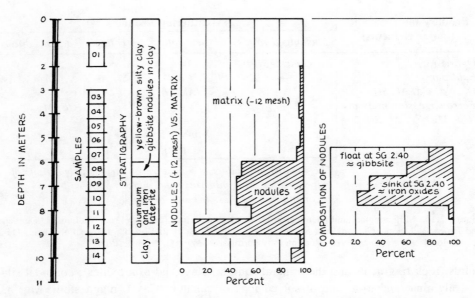

Figure 2.11. Some features of a bauxite deposit. (*Source:* W. H. Dennen and H. A. Norton. "Geology and geochemistry of bauxite deposits in the lower Amazon Basin." *Economic Geology* 72:82–89, 1972)

tendency of high density mineral particles to lag behind and accumulate in low spots in the channel or where water velocity is lower. Such accumulations are called placer deposits and such substances as gold, platinum metals, diamonds, cassiterite (tin ore), and ores of tantalum and columbium are to be found in them.

The placer gold deposits that triggered the great California Gold Rush of 1849 were formed in consequence of the extremely active erosion in the Sierra Nevada of California.

> Canyons several thousand feet in depth have been cut in an uplifted plateau, veritable trenches or sluice boxes, the grade of which is from 60 to 150 feet per mile. Stretches of wild gorges with polished bottoms alternate with stretches of less grade where shallow gravel accumulates. These canyons receive for long distances an abundant supply of gold, of all sizes, from older hill gravels or from decaying quartz veins. The result will be that but little gold will lodge in the gorges while extremely rich shallow gravel bars will accumulate in the convex stream curves [Fig. 2-12]. Gradient, volume, and load usually vary in the same stream so that deposition may be going on in one part of its valley and erosion in another. Continued corrosion of the stream-bed results in deepening the canyon and leaving the bars as elevated benches. The miners of 1849 first found these bars and worked them. In searching for the source of the gold they soon found a trail of metal leading up the gulches to great masses of older gravels on the hills, 2,000 to 3,000 feet above. These gravels were washed by the hydraulic method; and immense masses of tailings with a little gold were carried down to the rivers, totally overloading them. After the prohibition of hydraulic mining the streams gradually resumed active transportation. The whole gravel mass moved slowly downstream and a gradual reconcentration on the bed-rock took place. The tailings deposited became enriched and will ultimately be reworked" (Lindgren, 1933).

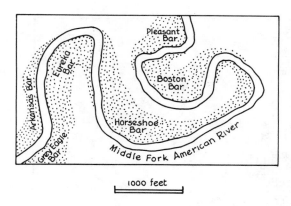

Figure 2.12. Gravel bars and gold placer deposits. (*Source:* Redrawn from W. Lindgren, *Mineral Deposits* New York: McGraw-Hill, 1919)

Weathering prepares materials for transport and the erosional agents carry the debris and dissolved substances away from the source area with varying degrees of selectivity and efficiency. The products of weathering are thus separated and deposited in different places or at different times. Deposition may be temporary, since most transportation is by fits and starts, but eventually the sorted debris will remain long enough in one place to be consolidated into a cohesive mass of rock. Lithification may occur by interpenetration of grains under compression as in a compaction shale, by minor recrystallization causing grains to interlock, or by the deposition of an interstitial cement from substances dissolved in groundwater. The result is a sedimentary rock, usually clearly layered or bedded, comprised dominantly of one or a few kinds of mineral grains of a common size, and having significant porosity. Less commonly, sedimentary rocks may be formed by the accumulation of vegetable or animal debris (coal, shell limestone) or the direct precipitation of dissolved substances by the evaporation of surface waters. Evaporite deposits are found worldwide and include economically important gypsum, the raw material of plaster, rock salt, borates, and various soluble salts of potassium. The distribution of some evaporite deposits in the conterminous United States is shown in Figure 2-13.

Sedimentary rocks may be the host for waterborne dissolved substances arising from igneous activity or abstracted by groundwater, or serve as resource materials in themselves. A useful classification of sedimentary rocks is provided in Table 2-3. Note that the various rocks, detrital, chemical, and organic sedimentary kinds, may have economic importance, especially at particular locations or if of high purity. Examples of detrital sedimentary rocks have been mentioned (placer deposits) and other economic deposits typical of sedimentary rocks that, when of adequate purity, may be used directly are:

Coal and related materials—fuel
Limestone—agricultural land plaster, lime, portland cement
Dolostone—agricultural land plaster
Gypsum—plaster
Quartz sandstone—glass-making
Clay and shale—tile and pottery

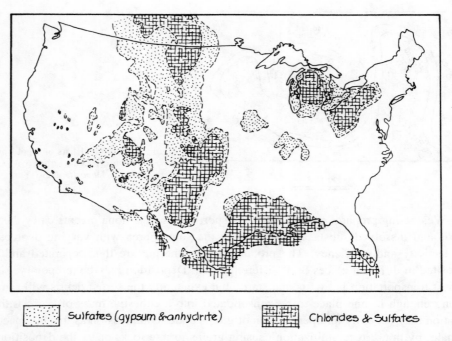

Figure 2.13. Evaporite deposits in the conterminous United States. (*Source:* Modified from *United States Geological Survey Professional Paper* 820, 1973)

Salt—table use and chemical industry
Phosphate rock—fertilizer

An example of a detrital sedimentary rock of commercial importance is the high-purity quartzose sandstone known as glass sand, the basic ingredient in the manufacture of glassware. Usable glass sands must be, or be readily upgraded to, a product of 95% or better quartz. Iron (except for beverage bottles!) is a particularly undesirable impurity since even at low concentrations it imparts a strong green or brown color to the product.

The very high purity of a typical glass sand has been attained by the reworking, often through many cycles of erosion and deposition, of pre-existing quartzose sandstones or quartzites. In the United States the most used rocks are the Oriskany Sandstone (Devonian) found in Pennsylvania and West Virginia and the St. Peter Sandstone (Ordovician) of Illinois and Missouri. The thickness of exploited deposits is usually a few tens of meters.

A temperature of 1,610° C is required to melt pure quartz, far too high for the manufacture of ordinary glasses. Consequently, various fluxes are added to the glass-making batch both to lower the melting point and incidentally to provide certain desirable qualities to the product. Table 2-4 gives the approximate compositions of some common glasses. The coloring of glass is accomplished by the addition of small amounts of particular chemicals to the melt; a partial list is given below:

Geologic Review

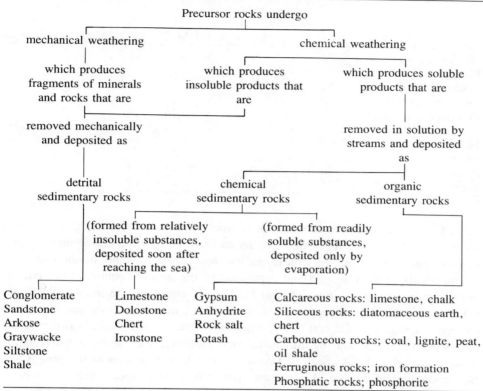

Table 2-3.
A Classification of Sedimentary Rocks

(Adapted from E. B. Branson, and W. A. Tarr, *Introduction to Geology*. New York: McGraw-Hill, 1935.)

    Ruby—gold
    Red—selenium
    Green—chromium, copper
    Blue—cobalt
    Yellow—cadmium, uranium
    Violet—manganese
    Brown—ferric oxide
    Opalescent—tin oxide, calcium fluoride

Iron formation, a principal source of iron ore, may be defined as a chemically precipitated sedimentary rock containing at least 15% iron. By far the greatest volume of this rock was deposited during the latter part of Lower Proterozoic time, about 1,000–2,000 million years ago. Most is an alternation of fine layers of chert and iron oxides, sulfides, or carbonates giving the rock a banded appearance, hence its acronym BIF for banded iron formation. Deposition was in shallow basins bor-

Table 2-4.
Approximate Composition of Some Common Glasses (percent)

	SiO_2	B_2O_3	Al_2O_3 + Fe_2O_3	CaO	MgO	Na_2O	MnO	PbO	K_2O
Window glass	72	—	1	8	4	14	—	—	—
Plate glass	72	—	—	11	2	14	—	—	—
Container glass	72	—	2	8	2	14	—	—	—
Container, amber	70	—	2	9	1	16	—	1	—
Container, dark green	67	—	5	8	1	15	1	2	—
Lead crystal	56	—	—	—	—	—	12	—	31
Fused silica	100	—	—	—	—	—	—	—	—
Vycor	96	3	—	—	—	—	—	—	—
Pyrex	80	12	3	—	—	4	—	—	—
Soda glass	70	1	3	5	4	17	—	—	—

dered by deeply weathered terrain of low relief. The typical stratigraphic section is a basement of Archean rocks followed by an early Proterozoic clastic unit, a basal volcanic and clastic unit, the iron formation, and finally an upper clastic unit.

The details of the transport and deposition of the iron are controversial, but the most popular concept is due to Cloud (1972). He correlates the timing of BIF development with the transformation of the Earth's atmosphere from one of no free oxygen (chemically reducing) to one containing free oxygen (oxidizing) during the time when oxygen-producing organisms were evolving. Weathering under reducing conditions during the Archean had released soluble ferrous iron to the oceans and the newly formed free oxygen reacted with it to form insoluble ferric iron compounds. When the ferrous iron had been scavenged, oxygen could accumulate as free molecules in the atmosphere. This great chemical event thus set the stage for the development of higher organisms and left a lasting inheritance of iron deposits (Fig. 2-14).

The red bed-copper association is a widespread and important type of ore deposit whose restriction to a particular sedimentary rock sequence places it with the primary sedimentary mineral deposits, although concentration of the ores by groundwater and fixation through bacterial action (Chapter 3) is perhaps important. The ores rarely, in fact, occur in red beds as implied by their name, but red-stained continental sandstones are a ubiquitous associate of the ore-bearing shales and sandstones. Carbonate rocks and evaporites are usually present in the postore sequence. Mineralization is uniquely confined to the first marine sediments laid down in an arid climate on a highly weathered basement. Representative deposits include the Central African Copperbelt in Zaire and Zambia, the Kupferschiefer, literally copper shale, of northern Europe, the White Pine District in Michigan, and the Coppermine River area of arctic Canada.

Phosphorite, which is used extensively in the manufacture of fertilizer, is another example of an important kind of mineral deposit formed as a chemically deposited sedimentary rock. As the name implies, phosphorite is a highly phosphatic rock

Figure 2.14. Deposits of Precambrian banded iron formation.

composed of at least 50% calcium phosphate in the form of the mineral apatite, a mineral also present as a principal component of bones and teeth.

Phosphorus is an essential constituent of all living things wherein it serves in key molecules such as DNA and ATP to store and transfer energy in metabolic processes.

Modern chemical fertilizers are designed to replace plant nutrients, principally nitrogen, potassium, and phosphorus, depleted by intensive agriculture. As mined, natural phosphates are too slowly soluble to be of great use so they are converted to readily soluble superphosphate by reaction with sulfuric acid.

Phosphorites are marine sedimentary rocks that have been formed in a somewhat restricted environment throughout the Phanerozoic. They are typically black, often pelletal, and closely associated with less phosphatic mudstones, chert, carbonate rocks, and greensands. They are formed by the precipitation of calcium phosphate on the sea floor either by a direct inorganic process or one mediated by microorganisms. Appropriate conditions for phosphate deposition are areas of upwelling of cold ocean waters into near-shore zones where precipitation of apatite is caused by (1) changes in temperature and the partial pressure of carbon dioxide, (2) reactions of phosphorus-rich waters with calcite, $CaCO_3$, or (3) biologic activity. Figure 2-15a shows the setting for phosphorite deposition together with an indication of the kinds of associated rocks to be anticipated.

Phosphate deposits are worldwide in occurrence, the more important being in the United States and northern Africa (Morocco, Tunis, Egypt). Most of the U.S. production is from deposits of Tertiary age in Florida. The largest reserve, however, is in the Phosphoria Formation (Permian) underlying about 250,000 km^2 in Utah, Idaho, Wyoming, Montana, and parts of Colorado and Nevada. This formation (Fig. 2-15b), 30–100 meters thick, contains at least 20 phosphate-rich beds aggregating about 22 meters.

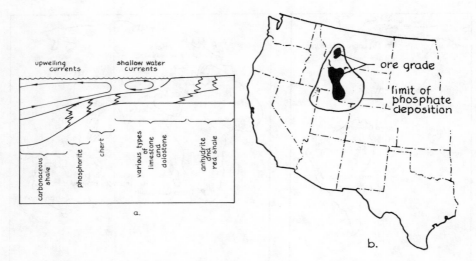

Figure 2.15. Phosphorite depositional setting and distribution of the Phosphoria Formation. (a) Depositional setting. (*Source:* After R. P. Sheldon, *United States Geological Survey Professional Paper* 313-B, 1963.) (b) Location of the Phosphoria Formation. (*Source:* After B. J. Skinner, *Earth Resources*. Englewood Cliffs, N.J.: Prentice-Hall, 1969)

METAMORPHISM AND METAMORPHIC ROCKS

Our active Earth continuously moves rock masses from one environment to another, and the rocks respond to the changes in temperature and pressure by rearrangement of their constituents into new mineral assemblages characteristic of the environment. The kind of changes that arise when deep-seated rocks are exposed at the surface were suggested in Figure 2-3, and reciprocal changes take place in rocks intruded by hot magma or carried to depth by burial or subduction where elevated temperature and pressure causes change (meta) of form (morph)—metamorphism.

The mineral assemblage that results from metamorphism will, of course, depend on the starting material as well as the level of temperature and pressure attained. Some terms used for metamorphic rocks that indicate their correlation with various starting materials are:

Starting Material	Metamorphic Rock Term
Volcanic rock	Metavolcanic; greenstone
Sandstone	Quartzite
Limestone	Marble
Dolostone	Dolomitic marble
Shale	Slate, phyllite (foliated)
	argillite, hornfels (nonfoliate)

Other terms are used to indicate the temperature level attained (grade) or the combined levels of temperature and pressure (facies). Metamorphic grade may be recognized by the presence of characteristic minerals that appear, if composition permits, in the sequence: chlorite (150° C), biotite, garnet, staurolite, kyanite, sillimanite (700° C). Metamorphic facies are described as characteristic mineral assemblages (Fig. 2-16).

The reorganizational processes that go on during prograde metamorphism are marked by a number of easily recognizable textural changes. Impurities are segregated into bands or isolated as well-formed crystals (metacrysts), often with consequent bleaching of the rock. Minerals of the same kind tend toward a uniform size, and directed pressure imposes an orientation of platy or bladed crystals causing a marked foliation of the rock.

Continued rise in temperature will eventually bring the rock to its melting point, thus generating a magma. Melting, however, does not occur simultaneously for the metamorphic rock mass but rather is stepwise. Rock melting and magma crystallization are both fractional and in a general way rock melting is the reverse of magma crystallization. The first products are usually quartz veins followed by simple pegmatites (quartz, feldspar, mica) lacking the magmatic suite of exotic minerals, and then by granitic intrusive bodies.

Metamorphic processes are doubly important because they may both cause concentration of useful substances into an ore deposit, and earlier formed deposits may be subjected to and modified by a later metamorphic event.

An intruding magma not only strongly heats the surrounding country rock, but often entails an exchange of material between the magma and its walls that contaminates the magma and generates an aureole of altered wall rock termed skarn or tactite. Workable deposits of metal-bearing minerals and such useful nonmetallics

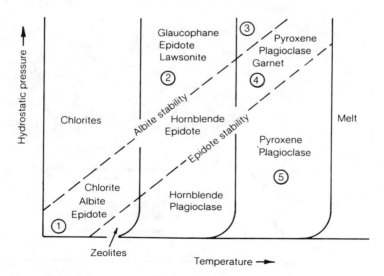

Figure 2.16. Metamorphic facies. Key: (1) Greenschist facies. (2) Blueschist facies. (3) Eclogite facies. (4) Granulite facies. (5) Pyroxene hornfels facies. (*Source:* From William H. Dennen and Bruce R. Moore, *Geology and Engineering*. Copyright © 1986 Wm. C. Brown Publishers, Dubuque, Iowa. All Rights Reserved. Reprinted by permission)

as graphite, garnet, and corundum are sometimes formed by metamorphism of this kind. The mineralized aureoles around the granitic plutons of the Cornwall district in England furnish a classic example of the formation of metalliferous deposits by contact metamorphism. This area was a principal source of tin in ancient times and tin mining of both lode and placer deposits continues today. In addition to tin, the Cornish mines or wheals have produced copper, arsenic, tungsten, lead, silver, uranium, iron, manganese, bismuth, nickel, and cobalt. Figure 2-17 shows the distribution of the plutons and their metamorphic aureoles, and the inset is a diagrammatic representation of the zonal arrangement of mineralization.

Regional metamorphism, involving the elevation of temperature and pressure in a large volume of the lithosphere, results in changes in the mass that may occasionally produce valuable minerals. Graphite from carbonaceous matter, kyanite from clay, and garnet from ferromagnesian minerals are examples that, if sufficiently concentrated, represent useful resources. Water contained in the rock promotes the crystalline reorganization and is essential to the formation of such products as talc, soapstone, pyrophyllite, and asbestos from appropriate precursor materials.

With increasing metamorphic intensity, water is progressively driven out of the zone of metamorphism carrying with it any dissolved substances. This water is a hydrothermal fluid with characteristics identical to one generated by igneous processes. Arguably, this is the origin of the copper deposited in the upper portions of ancient basaltic lavas in the Upper Peninsula of Michigan.

Large areas of metamorphosed terrain, for example, in the ancient shield areas of Canada, Fennoscandia, South America, Africa, India, and Australia, contain metal-

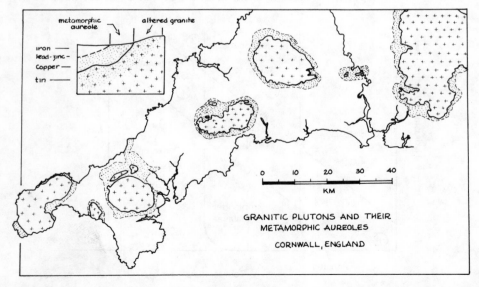

Figure 2.17. Granitic plutons and their metamorphic aureoles, Cornwall, England. (*Source:* Modified M. L. Jensen and A. M. Bateman, *Economic Mineral Deposits*. New York: Copyright © John Wiley & Sons, 1979. Reprinted with permission of John Wiley & Sons, Inc.)

liferous deposits, some of which must have existed prior to the metamorphism. Unfortunately, the criteria for the recognition of metamorphism of sulfide minerals is absent or questionable because of their ease of recrystallization into assemblages having all of the features of primary deposits. Broken Hill, Australia, used in the next chapter as an example of a deposit formed by the deposition of sulfides from solution in high temperature water, may also be interpreted as an older deposit that was rearranged during intense regional metamorphism.

References

Amstutz, G. C., Frenzel, G., Kluth, C., Moh, G., Wahnskund, A., Zimmerman, R. A., and Gorsey, A. El. 1982. *Ore Genesis.* New York: Springer-Verlag.

Bowen, N. L. 1928. *The Evolution of the Igneous Rocks.* Princeton, NJ: Princeton University Press.

Burnham, C. W., and Jahns, R. H. 1962. A method for determining solubility of water in silicate melts. *Am. Jour. Sci., vol.* 260.

Craig, J. R., Vaughan, D. J., and Skinner, B. 1988. *Resources of the Earth.* Englewood Cliffs, NJ: Prentice-Hall.

Cox, A., and Hart, R. B. 1986. *Plate Tectonics—How it Works.* London: Blackwell Scientific Publications.

Evans, A. M. ed. 1982. *Metallization Associated with Acid Magmatism.* New York: John Wiley & Sons.

Gass, I. G., ed. 1980. Volcanic processes in ore genesis. *Geol. Soc. London* Special Publication #7.

———. 1980. *An Introduction to Ore Geology.* New York: Elsevier.

Friedrich, G., and Herzig, P., eds. 1987. *Base Metal Sulfide Deposits in Sedimentary and Volcanic Environment.* New York: Springer-Verlag.

Goranson, R. W. 1931. Solubility of water in granitic magmas. *Am. Jour. Sci., vol.* 22.

Guilbert, J. M., and Park, C. F. Jr. 1986. *The Geology of Ore Deposits.* New York: W. H. Freeman.

Holmes, A. 1965. *Principles of Physical Geology.* New York: Ronald Press.

Jahns, R. H., and Burnham, C. W. 1969. Experimental studies of pegmatite genesis: I. A model for the derivation and crystallization of granitic pegmatites. *Economic Geology,* vol. 64.

Kennedy, G. C. 1955. Some aspects of water in rock melts. *Geol. Soc. America* Special Paper 62.

Lindgren, W. 1933. *Mineral Deposits,* 4th ed. New York: McGraw Hill.

Luth, W. C., and Tuttle, O. F. 1969. Hydrous vapor phase in equilibrium with granite and granite magmas. *Geol. Soc. America* Memoir 115.

Mitchell, A. H. G., and Garson, M. S. 1982. *Mineral Deposits and Global Tectonic Settings.* London: Academic Press.

Pettijohn, F. J. 1975. *Sedimentary Rocks,* 3rd ed. New York: Harper & Row.

Phillips, G. N., Myers, R. E., and Palmer, J. A. 1987. Problems with the placer model for Witwatersrand gold. *Geology,* vol. 15.

Press, F., and Seiver, R. 1982. *Earth,* 3rd ed. San Francisco: W. H. Freeman.

Sawkins, F. J. 1984. *Metal Deposits in Relation to Plate Tectonics*. New York: Springer-Verlag.

Tarling, D. H., ed. 1981. *Economic Geology and Geotectonics*. New York: John Wiley & Sons.

Walker, W., ed. 1976. *Metallogeny and Global Tectonics*. Stroudsburg, PA: Dowden, Hutchinson & Ross.

Chapter 3
Groundwater and Ore Solutions

WATER UNDERGROUND

Almost without exception, mineral deposits are formed by some process related to the physical or chemical action of water. It is thus only appropriate that some special attention be given to this critical but fugitive substance.

The crust of the Earth is saturated with fluids filling its cracks and intergranular spaces from an upper surface, the *water table,* to such depths that it has been completely squeezed out by rock pressure or immobilized by chemical reactions. In volume, the crust is estimated to contain about one-half as much water as is found in the oceans. Thus it may be said that the Earth's crust is made of a solid but more or less porous framework filled with mobile, infinitely deformable fluids. The dominant fluid is water, but there may also be other liquids or gases. The water may be *juvenile*—added from below by igneous activity or metamorphism, *connate*—trapped in sediments as they are deposited, or *meteoric*—water added by infiltration from the surface. Water also exists in rocks and soils between the water table and the earth's surface, but as films adsorbed on grains rather than as a mobile medium.

The realm of underground water may be usefully divided into three distinct but interconnected zones, each with characteristic hydraulic settings. In the uppermost or *phreatic* zone above the water table, the amount of moisture is controlled by the amount of water infiltrating down from the surface, its rate of downward transport, and losses to the zone of saturation below the water table, and by evaporation. The amount of water that can be held in this zone without movement is termed the *field capacity*.

Water in connected openings below the water table interacts with the phreatic zone through a *capillary fringe* and outcrops as surface waters such as lakes or rivers. This water can move more or less rapidly under the influence of a hydraulic head depending on the size and connectivity of the openings. It is this mobile groundwater, exceptionally extending downward to several thousand meters but usually much shallower, that is exhaustively treated in texts on hydrology and groundwater geology.

With increasing depth, the continuity of openings is lost due to increasing lithostatic pressure and the water in less compressible rock bodies becomes isolated as connate water. Juvenile water may also be newly released in this deep zone in consequence of magma crystallization or dehydration of original rock material by metamorphism. Although small in amount, water in the deep crust is a significant contributor to the mechanical properties of the rocks, their chemical modification, and a principal means of chemical transport.

Groundwater is far from chemically pure. Water is an excellent solvent and becomes increasingly reactive as its temperature is increased with depth. Being in intimate contact with the rock matrix and moving slowly, groundwater may be expected to attain chemical equilibrium with those minerals with which it is in contact, dissolving and depositing mineral matter as it moves. A means for the acquisition, transport, and deposition of substances within the Earth through the agency of groundwater is thus present.

POROSITY AND PERMEABILITY

The amount of open space in soils and rocks is termed *porosity*. In a physical way, the openings might, for example, be intergranular pores, cracks or joints, vesicles in lava, or cavernous openings in limestone. Porosity may be given as the percent of void space in a solid or as the *void ratio,* e, defined as

$$e = n/n - 1$$

where n is the porosity as a decimal fraction of unity, that is, 30% = 0.30. When connected, rock openings allow the movement of water through them. This imparts the very important property of hydraulic conductivity or *permeability* to a soil or rock. Both the property and permeability of rocks and soils vary over an enormous range and are not necessarily comparable. Freshly deposited clay, for example, might incorporate 45 volume % of water between the clay particles (i.e., have a porosity of 45%), but have a negligible permeability because the water is not free to move. Alternatively, the low porosity of a granite might be concentrated in a single crack resulting in significant permeability of the rock.

The porosity of granular rocks is a function of the size, size distribution, shape, and arrangement or packing of the grains (Fig. 3-1). Most classic sedimentary rocks have porosities between 10 and 45%, although some poorly consolidated sediments may have porosities as high as 80% (e > 1) and recently deposited muds may hold up to 90% water by volume. The porosities of sediments decrease as they become consolidated into rock, and it should be noted that the water expressed from them must go somewhere. Igneous rocks are at the other end of the porosity scale with unfractured granite, gabbro, and obsidian having essentially zero porosity. In general, a porosity of less than 5% is considered low, 5% to 15% as medium, and over 15% is regarded as high.

The porosity of rocks ascribable to cracks is particularly important because the mechanisms of crack generation are often such that fracturing tends to be localized

Groundwater and Ore Solutions

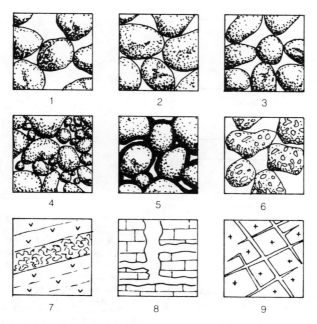

Figure 3.1. Porosity in geologic materials. Key: (1) Open packing, porosity higher. (2) Close packing, porosity lower. (3) Good sorting, porosity higher. (4) Poor sorting, porosity lower. (5) Cementation, porosity reduced. (6) Porous grains, porosity increased. (7) Porous zone between lava flows. (8) Solution joints in limesotine. (9) Fractures in crystalline rock. (*Source:* From William H. Dennen and Bruce R. Moore, *Geology and Engineering.* Copyright © Wm. C. Brown Publishers, Dubuque, Iowa. All rights reserved. Reprinted by permission)

into extensive tabular zones. Such a concentration of cracks thus results in a conduit through which enhanced water movement can take place.

Permeability is a measure of the ease with which fluids can pass through a porous medium such as soil or rock. A rock that carries significant amounts of water underground is called an *aquifer* (AQUI, water; transFER). In contrast, a rock of low permeability is termed an *aquaclude* (AQUA, water; exCLUDE).

Permeability is related to both the size and connectivity of pore passages and cracks. Pore passage diameters vary roughly with particle diameters in granular rocks and it is common practice to relate the permeability of such rocks to grain size. The approximation, following Hazen (1893), is that

$$K = 100(D_{10})^2$$

where K is the permeability in cm/s and D_{10} is the diameter of particles of 10% finer weight as discussed in Chapter 1. The concept is that finer material occupies the interstices between the larger grains and permeability will thus be a function of the amount of finer fraction. Figure 3-2 shows the permeability of some common sediments; the regularity of the changes in permeability with D_{10} should be noted.

The change in the permeability of rocks in consequence of fracturing is marked and results in increases of 10^2 and 10^9 in their ability to transmit water.

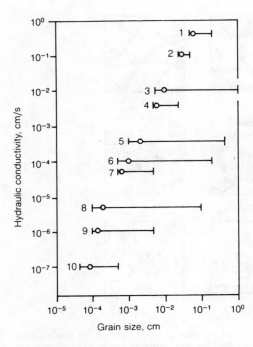

Figure 3.2. Permeability of some common sediments. Key: (1) Uniform coarse sand. (2) Uniform medium sand. (3) Well-graded sand. (4) Uniform fine sand. (5) Well-graded silty sand and gravel. (6) Silty sand. (7) Uniform silt. (8) Sandy clay. (9) Silty clay. (10) Clay (30–90% clay sizes) D_{10} value indicated by open circle. (*Source:* Drawn from data in B. K. Hough, *Basic Soil Engineering* 2/e. New York: Copyright © Ronald Press Co., John Wiley & Sons, Inc.)

Water is essentially incompressible and infinitely deformable so pressure applied at any point on a water mass causes it to deform and flow if not constrained. The ease of movement of water through porous materials, however, varies widely with the grain size, packing, and degree of cementation of the rock. The rate of movement of water underground is controlled by fluid potential difference and the size and connectivity of the openings in the host rock. The relations are given by the empirical law of Darcy (1856), which may be stated as

$$Q/A = k\Phi/n$$

where Q is the volume of fluid per unit time passing through an area A in the direction of the pressure gradient Φ; k is the permeability related to properties of the rock and rock openings including friction; and n the coefficient of viscosity of the fluid. It may be seen (Fig. 3-3) that the quantity of water passing through a given cross-section is strongly dependent upon k. For example, the water yield per unit time between clean sand ($k \approx 10^{-3}$) and silt ($k \approx 10^{-6}$) under the same pressure differential is 1,000 to 1. Water transport through fractured rock may be a million or more times greater than through the same rock without fractures. The coefficient of viscosity of water is strongly dependent upon temperature, decreasing exponentially from 1.05 centipoise at 18° C to 0.28 cp at 100° C. Q/A will thus change by a factor of 3.75 when water experiences this temperature change.

Because liquids interact physically with solids as seen in surface tension phenomena, the theoretical permeability of a material must be corrected downward to accommodate these frictional effects. The effective velocity for water movement through

Groundwater and Ore Solutions

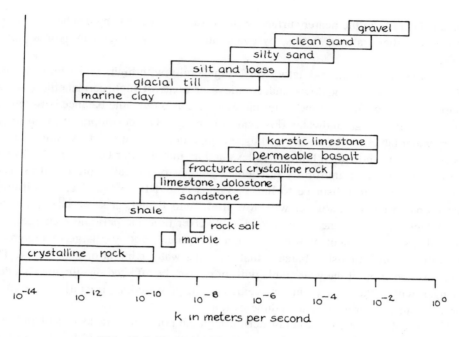

Figure 3.3. Permeability of some sediments and rocks.

rocks is usually very small, but is never zero. Even rocks with such low permeabilities as to entrap water for geologic times (connate water) show by the excessive saltiness of the remaining water that some has been lost by osmosis.

SHALLOW GROUNDWATER

Groundwater infiltrates through the phreatic zone and circulates, albeit very slowly, through the cracks and connected pores of soil and rock below the water table. With depth, these openings are gradually squeezed shut and when of capillary or smaller size the attraction of the channel walls effectively stops water movement. Above this depth of a few thousand meters or less, depending on the rock, circulation results from differences in head arising from the irregular surface of the water table. This may be called the zone of shallow groundwater.

Water movement in the phreatic and shallow groundwater zones is dominantly in consequence of the force of gravity acting on the water mass. Surface water infiltrates through the phreatic zone to the water table and then moves through the shallow groundwater zone from points of higher to lower pressure.

Addition of water to these zones is principally through recharge from rainfall and surface water, although recharge from below from magmatically or metamorphically generated waters must occur. Such water rises to join and blend with downward-moving water from the surface. Observations in geothermal fields suggest that this rise of heated water is not on a broad front, but rather in a column or plume that spreads laterally in the nearer surface regions. Loss of water is by discharge through springs or other leakages, wells, by evaporation, and by fixation in hydrated minerals.

The water table is an undulatory surface generally with highs under hills or areas of low permeability and lows under valleys or areas of high permeability. In the short term, the position of the water table represents a dynamic balance between the rates of recharge, groundwater flow, and discharge. The development of an undulatory water table may be understood by consideration of Figure 3-4. Assume a cross-section, line 1, showing a hill and valleys and an initial water table, line 2. Assume the area to receive a uniformly distributed amount of rainfall represented as infiltrating by a uniform distance to line 3. Points A and B on line 3 are at different elevations so the groundwater will flow from A to B and, at some later time, the water table will be at the position of line 4. Over time the periodic recharge from rainfall, slow movement of water through the ground, and discharge to streams in the valleys will establish a balance that fixes the water table at or near line 4. The long-term position of the water table will, of course, be affected by climatic changes, changes within the rock (solution, cementation, compaction, fracturing), uplift of the land surface, erosion, and sedimentation.

The percolation rate of groundwater through the pores and cracks of soil and rock is much less than the recharge from the surface so a water mound builds up under topographic highs. This mound forces the subjacent water mass to move toward areas of lower pressure, typically at topographic lows, and a pattern of flow lines reminiscent of electric or magnetic field lines develops (Fig. 3-5). The subsurface flow of the groundwater is along downward-looping paths from areas of recharge to those of discharge and the region becomes divided into alternating cells of recharge and discharge.

Figure 3-6 shows how the underground water flow might appear in three dimensions. The topography (dashed) has been removed and the undulating water table shown as a contoured surface. Water movement on and below this surface is indicated by arrowed flow lines that diverge from mounds and converge to trenches. Assuming the saturated ground to be equally permeable in all directions, the flow lines will be perpendicular to equipotential surfaces (surfaces of equal pore pressure) below the water table. These surfaces are indicated by dotted lines on the face of the block. Flow from any recharge area on the water table surface, say from A to B, is within a flow channel bounded by flow surfaces, as shown in the inset, whose shape is dictated by the position of equipotential surfaces.

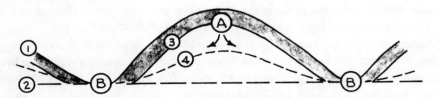

Figure 3.4. Development of a water table (see text for details.)

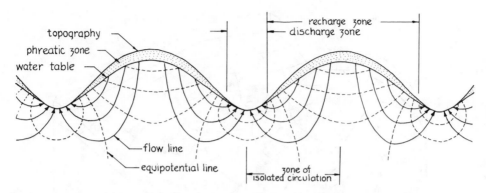

Figure 3.5. Cross section of regional groundwater flow through homogenous material. (*Source:* Redrawn after M. K. Hubbert, "The theory of groundwater motion," *Journal of Geology*, 48:785–944. By permission of the University of Chicago Press)

The above discussion assumes the permeability of the subsurface materials to be uniform in all directions. It is unrealistic, however, to assume homogeneity of subsurface conditions over any significant volume. Real rocks will show more or less marked differences in their permeability from point to point. Flow in materials of differential permeability will be in the direction of the resultant of the greatest and least permeability present as shown in Figure 3-7. The channelizing effect of highly permeable zones such as aquifers, open fractures and joints, or solution passages should be remarked. Consider, for example, the effects of a more permeable layer or fissure on the flow paths or groundwater as shown diagrammatically in Figure 3-8b, c. These high-permeability shortcuts divert the flow within the zone and the shape of adjacent zones is modified in compensation. Other phenomena arising from differential permeability in the subsurface are related to the confinement of water by aquacludes. For example, in Figure 3-9, water in well A will rise to the potentiometric surface and in well B will fountain as an artesian well. Related geometries and pressure differences cause oil to gush out of oil wells.

DEEP GROUNDWATER

Deep groundwater may be defined as that portion of the water underground whose motion is not sensibly affected by the differences in elevation of the water table that are the drive for the continuous movement of shallow groundwater. Rather, the motion of deep groundwater is aperiodic and event-related.

The evidence for the presence of groundwater in the deeper regions of the Earth's crust is overwhelming. Direct evidence is possible in the waters trapped with oil or gas accumulations or present in geothermal fields; indirect evidence is seen in extensive zones of rock alteration, epigenetic mineralization in veins and disseminations, and inferential evidence comes from the theory and observations of rock mechanics, igneous petrology, and metamorphic petrology.

Water is intimately involved in the Earth's rock-forming processes. It may be

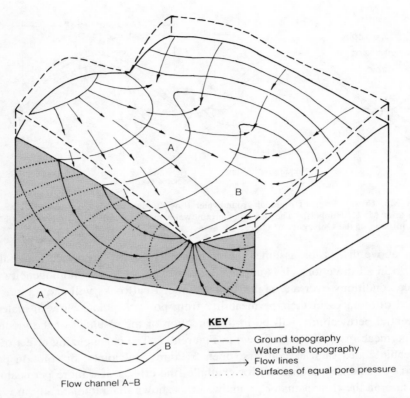

Figure 3.6. Groundwater flow. (*Source:* From William H. Dennen and Bruce R. Moore, *Geology and Engineering*. Copyright © 1986 Wm. C. Brown Publishers, Dubuque, Iowa. All rights reserved. Reprinted by permission)

exsolved from magma as the mass cools, sediments trap water in their pores as they are deposited, and pre-existing rocks are dehydrated by metamorphism. Further, there must be changes in the water content of a rock mass as the rocks are modified by processes that change their porosity or proportions of hydrous and hydrated minerals. Water thus must be both present and move in the crust of the Earth below the shallow zone of free circulation. In this region of deep groundwater, however, we are faced with abundant evidence of water movement but have little quantitative knowledge of the mechanism.

Connate waters originate as the pore water of sediments, trapped therein by burial and carried to depth by tectonic processes. Originally the water content of sediments may be very high—45% for pelitic and 30% for psammitic sediments—but drops to a few percent as the pore space is reduced by lithostatic pressure and cementation. These processes thus initiate the movement of water from the rock.

The exsolution of water from water-saturated magma as it rises into regions of lowered pressure is described in Chapter 2. A further mechanism whereby water is released is the so-called second boiling. As magma solidifies, the minerals that crystallize are mainly anhydrous or use very little water in their structure. In consequence, the initial water content of the magma is little changed in amount as crys-

Groundwater and Ore Solutions

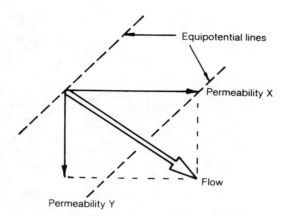

Figure 3.7. Groundwater flow in an antisotropic medium. (*Source:* From William H. Dennen and Bruce R. Moore, *Geology and Engineering*. Copyright © 1986 Wm. C. Brown Publishers, Dubuque, Iowa. All rights reserved. Reprinted by permission)

tallization proceeds, but is confined to a continually shrinking volume. Eventually ebullition of water takes place. For example, in Figure 3-10 a magma at 900° C under a pressure of 1 kbar might initially contain 4 weight % water, point A. The magma is not saturated with water at this point but will become so when it cools to 760° C, point B. Further cooling with consequent crystallization of anydrous solid phases reduces the volume of the magmatic liquid and increases the concentration of water. This continues to 670° C, point C, when the vapor tension reaches the

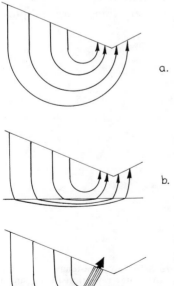

Figure 3.8. Effect of differing permeability on groundwater flow. (a) Groundwater flow lines from recharge to discharge zones through homogenous material. (b) Higher permeability material below concentrates the flow. (c) The effect of a fissure is to divert and concentrate the flow.

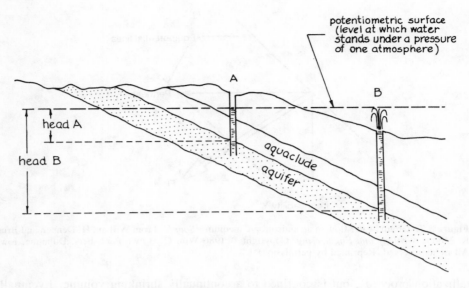

Figure 3.9. Conditions for artesian flow.

confining pressure of 1 kbar. At this point, boiling begins and continues as long as crystallization proceeds.

Metamorphism typically reduces the water content of the protoliths, the amount of water being released at the reaction site being a function of the original rock composition and the intensity of the metamorphic reaction. The release of water is probably in pulses whose sharpness reflects the bulk composition of the original rock because dehydration reactions are mineral specific at a given temperature and pres-

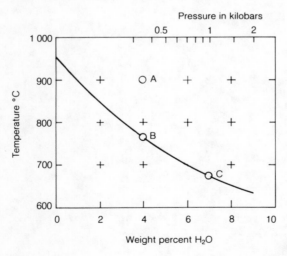

KEY

A   Undersaturated magma
B   Saturated magma
C   Second boiling when vapor pressure equals assumed confining pressure

Figure 3.10. Pressure-temperature relations for water retention in a granitic magma. (*Source:* From William H. Dennen and Bruce R. Moore, *Geology and Engineering*. Copyright © 1986 Wm. C. Brown Publishers, Dubuque, Iowa. All rights reserved. Reprinted by permission)

sure. Figure 3-11 shows the postulated relations between water content and temperature for basaltic and pelitic sedimentary rock undergoing progressive metamorphism with burial in a region whose geothermal gradient is 20° km^{-1}. Note that initially the basalt is hydrated, after which progressive dehydration takes place whereas dehydration is continuous for a pelitic sediment under the same conditions. In both instances metamorphism to middle grades results in the release of several weight percent water.

Water released by reactions and processes in the deep crust would have little importance in the formation of hydrothermal mineral deposits were it not for the fact that water moving away from its source becomes focused and channelized. Channelization is accomplished either by its movement through rock layers of higher permeability or along fractures or fracture zones induced by mechanical failure.

The strength of rocks is increased by increasing confining pressure but lowered by a rise in temperature, strain rate, fluid content, and pore pressure. Temperature and total confining pressure are depth-related independent variables, whereas pore pressure, differential stress, and strain rate are usually interdependent.

Rock masses within the earth are continuously subject to load stresses and aperiodically stressed by magmatic intrusion and tectonism. Under sufficient differential stress, the rocks will fail by rupture, folding, or flowage, or some combination. The presence of water in buried rocks, even in tiny amounts, markedly reduces their ultimate strength. Water thus promotes those deformational events that provide those openings needed for its movement.

The uncorrected pore pressure of water at any depth is readily determined because

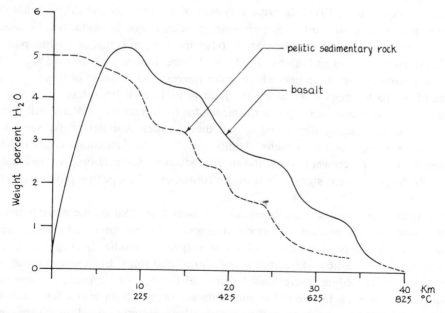

Figure 3.11. Loss of water with progressive metamorphism. (*Source:* Redrawn from W. S. Fyfe, N. J. Price, and A. B. Thompson, *Fluids in the Earth's Crust.* Amsterdam: Elsevier, 1978)

the pressure of a 1-centimeter cube of water on its lower face is 1 gram/cm^2 and that of a 1-meter cube is 1 tonne per square meter. For a head of 100 meters, the pressure is 100 tonnes per square meter. Unconstrained water at a depth of 5 kilometers is under a hydraulic pressure of 5,000 t/m^2, approximately 0.5 kilobars. Greater accuracy in such calculations requires adjustment of the density of water for elevated temperatures, a decrease of about 4% for each 100° C rise, for dissolved substances, and the pressure increment due to lithostatic pressure.

Hydrostatic pressures well above those calculated for a free water column are regularly attained in the Earth's crust because the movement of water is impeded by friction and surface tension. Since it cannot flow freely, the pore water assumes some of the load of the heavier rock column, and the lower the permeability the higher are the hydrostatic pressures that are attained. Water may thus be the causative agent for rock failure in those circumstances that its pore pressure equals or exceeds the lithostatic pressure.

The strength of rocks at depth can thus be exceeded by external forces—tectonic forces, igneous intrusion—or internal forces arising from the presence of incompressible water within a rock body. At depths of about 1.5 kilometers the pore pressure of water approaches that of the lithostatic pressure and further compression, differential stress, or reduction in pore volume must result in failure by rupturing. Indeed, hydraulic fracturing by water or magma is perhaps the most important single mechanism of deformation operative in the upper crust of the Earth. Its effects are seen on a small scale in the tiny fractures formed in cataclastic deformation and fracture cleavage, on an intermediate scale in the generation of joints and veins, and on a large scale by concordant and discordant igneous intrusions.

Fyfe and coworkers (1978) describe a system of water "sills and dikes" whereby the pore pressures of upward-migrating water ponded below an aquaclude becomes sufficient to lift and fracture the barrier. Another mechanism is that of seismic pumping described by Sibson and others (1975). Prefailure dilatancy of stressed rock creates low-pressure pore space into which water moves, and collapse of this space on failure of the rock forces it out along the fault or associated fractures.

Magma intrusion provides a powerful means for rock fracturing. Water will move into or out of the magma depending on whether the melt is under or oversaturated with water at the existing conditions. Additionally, igneous intrusion creates a thermal regime favoring convective motion of groundwater. Such motion, for example, is thought to be of great significance in the formation of copper porphyry deposits (Fig. 3-12).

Rock failure, by whatever mechanism, is conservative; that is, the deformational strain is minimal. The resultant deformation as seen in the microscopic to megascopic displacement of one portion of the mass with respect to another, as represented by joints, cracks, cataclastic and fracture zones, faults, and folds. Even when continuity is preserved as in folding, there must be internal flowage or slippage. A most important feature of rock failure is the concentration of the strain into a few more or less open and extensive fractures having a particular orientation fixed by the strength properties of the rock and the orientation of the stress field. Fractures increase the

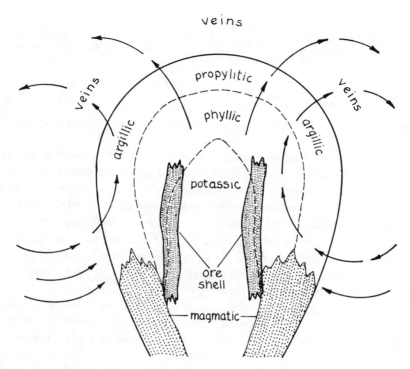

Figure 3.12. Schematic cross section through an idealized porphyry copper deposit showing the location of characteristic alteration and groundwater circulation. (*Source:* Redrawn after J. D. Lowell and J. M. Guilbert, *Economic Geology,* 65, 1970)

permeability of rock by orders of magnitude, and their nature requires the flow through them to be channelized.* Channelways of this kind often intersect the earth's surface as linear features that can be recognized by geologic study or remote sensing techniques (see Chapter 4) and thus serve as prospecting targets.

Migration routes of deep groundwater may only be inferred, but are undoubtedly similar in general to permeability-controlled flows at higher levels. The general tendency of the flow will be from points of higher to lower pressures and, for the most part, will have an upward component. Small differences in permeability will result in significantly different velocities and the flow pattern may be greatly influenced by rather minor structural features. The amount of water moving through a given cross-section in a stack of sedimentary beds of different permeability will, for example, be as suggested by the vertical section of Figure 3-13. Taking the lowest value of k (10^{-8}) as unity, 1,000 times more water moves through an horizon whose k is 10^{-5}. Reduction in either the permeability or cross-sectional area along the movement direction represents a flow constriction with consequent trapping or ponding of the fluid.

---

*Chemical studies of mylonite (crushed and healed zones) in granite have shown that as much as 400 volumes of water per volume of rock have passed through them (personal communication, W. H. Blackburn, 1988).

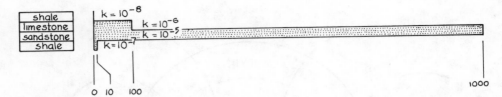

Figure 3.13. Relative flow distance in equal time through rocks of different permeability.

Similar considerations lead to the focusing of flow in plan view. Figure 3-14a is a contour map of the top of the Knox Dolomite in Kentucky centered on the Lexington (or Jessamine) Dome of the Cincinnati Arch. Arrows show the postulated routes of groundwater movement. Connate waters squeezed out of the rocks in the deeper parts of the basins to the east and west migrate laterally and upward toward the arch through the more permeable horizons. The unconformity at the top of the Cambro-Ordovician Knox Dolomite is thought to be such an horizon due to presence therein of dead oil and scattered sulfide minerals, both marking the passage of fluids. The Knox unconformity is a somewhat irregular surface and channelization of fluids should occur into and along the antiforms as shown in Figure 3-14b. Interruption of flow by a reversal of dip or locally decreased permeability results in ponding and possible deposition. Faults associated with the Kentucky River Fault Zone are present near the crest of the dome and penetrate to and beyond the top of the Knox Dolomite. If open and thus highly permeable, they provide escape routes for subterranean waters and, in this instance, the localization of fault-controlled veins of (noneconomic) barite, sphalerite, and galena.

Other postulated patterns of water movement in the deep and shallow groundwater zones are shown in Figure 3-15.

ORE SOLUTION AND DEPOSITION

Natural waters are solutions of various substances and concentrations. The kind of dissolved material depends on the original source of the water and its solute modification by reactions with and solution of the solids with which it is in contact. The leaching of mineral matter by groundwater is a function of the solubility of the solid phase, which is, in turn, dependent on the composition of the solvent, temperature, pressure, and pH. Thermal waters in various parts of the world have been shown to contain elevated concentrations of metallic elements in solution or depositing as minerals (see Table 3-1 for a partial listing).

Mineral solubilities in pure water are generally low, for example, for galena at saturation only 0.86 grams of lead in each cubic meter of water, and the transport

Figure 3.14. Channelization of groundwater. (a) Possible channelization routes (arrows) on the top of the Knox Dolomite in Kentucky; structure contours in feet. (*Source:* W. H. McGuire and P. Howell, *Oil and Gas Possibilities of the Cambrian and Lower Ordovician in Kentucky*. Lexington: Spindletop Research Center, 1963) (b) Channelization of groundwater by flexures.

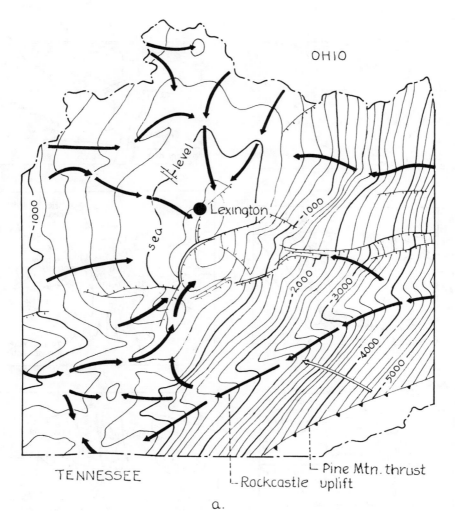
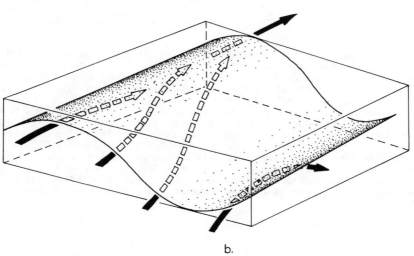

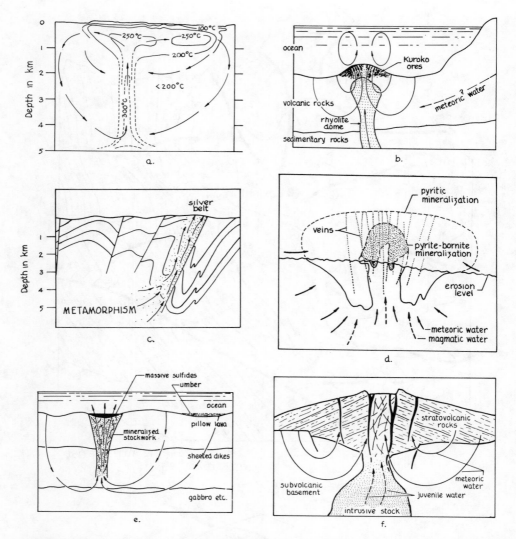

Figure 3.15. Examples of postulated deep groundwater flow. (a) High temperature water flow and temperature contours beneath the Waireki, N.Z., geothermal field. (*Source:* After J. W. Elder in "Terrestrial Heat Flow," *AGU Geophysics Monograph* No. 8, p211–239, 1965) (b) Conditions to form Kuroko ores. (*Source:* After O. Ohmoto and R. O. Rye, *Economic Geology,* 69:947–953, 1974) (c) Ore solutions from metamorphism. (*Source:* After D. L. Leach, G. P. Landis, and A. H. Hofstra, "Metamorphic origin of the Coeur d'Alene precious metal veins in the Belt Basin, Idaho, and Montana," *Geology,* vol. 16, no. 2) (d) Sulfide mineralization, El Salvador porphyry copper deposit, Chile. (*Source:* After L. B. Gustafson and J. P. Hunt, *Economic Geology,* vol. 70, 1975) (e) Postulated hydrothermal convection system, Troodis Massif, Cyprus. (*Source:* After E. T. C. Spooner, H. J. Chapman, and J. D. Smewing, "Strontium isotropic contamination and oxidation during ocean floor metamorphism of the ophiolitic rocks of the Troodos Massif, Cyprus. *Geochim. et Cosmochim. Acta,* 41:873–890, 1977) (f) Geologic setting for epithermal precious and base metal vein deposits. (*Source:* Adapted with permission from *The Geology of Ore Deposits,* by J. M. Guilbert and C. F. Park, Jr. Copyright © 1986 New York: W. H. Freeman and Co.)

Table 3-1.
Elements and Minerals

Elements Reported in Geothermal Waters		
antimony	lead	sulfur
arsenic	manganese	thallium
beryllium	mercury	tungsten
copper	nickel	zinc
gold	phosphorus	
iron	silver	

Minerals Deposited at the Surface from Geothermal Waters		
adularia	cinnabar	quartz
alunite	galena	realgar
barite	gypsum	silver
borax	kaolinite	sphalerite
calcite	pyrite	stibnite
chalcopyrite	pyrrhotite	sulfur

of most elements as simple dissolved solids requires unrealistic volumes of fluids. At the other extreme, the solubility of halite, NaCl, is 360 kilograms per cubic meter. Fortunately, the solubility of some elements, particularly metals, may be enhanced by orders of magnitude by their tendency to form relatively soluble, charged polyatomic groups or *complex ions* with such elements as chlorine or sulfur. Among the many complex ions that have been suggested or studied are metal sulfides, polysulfides, and thiosulfides, sulfates and thiosulfates, chlorides, carbonates, and hydroxides. No evidence of the particular soluble species remains after deposition since the deposited phase depends upon temperature, pressure, and concentration rather than the complex by which the metal is transported.

There are long-running arguments in geology as to the details of the nature and origin of oreforming solutions. However, their nature cannot be far different from that of groundwater in general and their ultimate origin as juvenile, connate, or meteoric water solutions cannot usually be distinguished after the mineralizing episode except probablistically. All water at depth must be hot (i.e., hydrothermal) and the real problems are those of solution, transportation, and deposition. These lie in the kinetics and species involved and their response to changes in temperature, pressure, pH, and concentration including mutual reactions and synergistic effects.

Although the transport and deposition of metals in oreforming solutions involves chemical and physical processes that are poorly understood, the subject is one of intense investigation. It appears that a majority of metals are carried as metal-chloride complexes in high-temperature chlorine-rich waters as evidenced by the abundance of chlorides in fluid inclusions in ore and gangue minerals. Under these conditions the hydrogen present is bonded to chlorine and the acidity of the solution, which depends on free hydrogen ions, is not far from neutrality. With lowered temperature, oxidation-reduction reactions, or fluid mixing hydrogen is freed from its association with chlorine, the increasingly acidic solutions react with and alter their walls and form hydrogen compounds in solution such as H_2S, HS^-, or $H_2CO_3^-$. In turn, these compounds react with the metal-chloride complexes to deposit metal sulfides and release more hydrogen chloride (hydrochloric acid). For example:

$$PbCl_2 + H_2S = PbS + 2HCl$$

It is easy to see that many complex reaction sequences may take place as waters migrate and adjust their dissolved load to changing conditions. The character of the solution may change in time resulting in cracks filled from the walls inward by successively different minerals, mineral assemblages may change with depth, simultaneous solution of one mineral and deposition of another in its place may result in volume for volume replacement of wall rock or earlier-formed vein minerals, or deposition may be restricted to a particular kind of host rock. Further, earlier veins may be more brittle than their wall rocks and crackle under stress to provide reopened channels for another generation of ore solutions.

The acquisition, transport, and deposition of mineral matter by water underground to sufficient levels to be considered ore calls for very special geologic conditions. First, all parts of the system must be simultaneously operative, and second, the concentrations in solution and rate of transport must be adequate. Unless a source, open transport route, and mechanism for deposition are simultaneously present, no orebody can be formed. Generation of hydrothermal solutions by igneous and metamorphic processes or activation of connate water by tectonic movements must be coincident in time with the opening of adequate channelways or presence of zones of high hydraulic conductivity in order that ore solutions can be transported and concentrated into a workable deposit. The probability is high that the ore-forming process will be completed in a relatively short time since it is unlikely for geologic conditions to remain favorable for more than a few thousands of years.

It is no accident that ore deposits most often occur in orogenic regions—mountain building and its associated igneous and metamorphic activity provide the necessary conditions for mineralization. Magma is a source of ore elements and its intrusion opens the essential transport route; connate water is liberated from rock masses by tectonically-generated stress; and water with its dissolved load is liberated from metamorphosing rocks.

Low concentrations and slow transport rates will be inadequate to bring sufficient material to a depositional site to constitute ore. For example, barite-saturated groundwater contains about 5 grams per cubic meter at 100° C and about 2 grams at 18° C. If an upward-moving saturated solution cooled rapidly from 100° C to surface temperature, it would deposit about 3 grams of barite from each cubic meter of water. If the groundwater velocity was 10^{-6} meters per second, a reasonable value for sandstone, only about 100 grams of barite would be deposited annually, or about 1 tonne in 1,000 years. Should, however, the groundwater velocity be 10^{-2} meters per second because of easy movement through a fracture, the same conditions would yield a deposit of 10,000 tonnes in the same time.

The physical conditions at the depositional site must be such as to provide adequate space for deposition or chemical conditions be such that the ore minerals can make space for themselves by replacement of the country rock, the modified solution moving away from the depositional site.

Groundwater rising from deep to shallow to surface regimes must undergo considerable changes in its environment, which will be particularly marked at the interfaces between zones. Deep groundwater will usually be hotter, more chloride-rich, and carry a larger dissolved load than will waters of shallow circulation. Deposition of solids from the deep waters will be triggered by dilution and decrease in temperature when it encounters shallow groundwater.

There is an extreme contrast in chemical conditions across the water table that results in active deposition at this level. Below the water table the openings are water-filled, pH is low (reducing), the partial pressure of oxygen is low, and anaerobic bacteria may flourish. The situation is reversed above the table with air as well as water in the pores, thus leading to oxidizing conditions and abundant aerobic organisms.

Groundwater chemistry above the water table is dominated by that of the atmosphere with dissolved oxygen and carbon dioxide as the principal reactants. The respiration and decay of organisms in the phreatic zone leads to an enrichment of CO_2 in the soil gas. Carbon dioxide also is more soluble in water than the more abundant atmospheric gases and, with water, yields carbonic acid:

$$CO_2 + H_2O = H_2CO_3$$

In solution, this acid dissociates to HCO_3^- and to a lesser extent to the carbonate radical, CO_3^{2-}

$$H_2CO_3 = H^+ + HCO_3^- = 2H^+ + CO_3^{2-}$$

HYDROTHERMAL ORE DEPOSITS

Deposition of minerals from water solution will take place under appropriate physico-chemical conditions and be generally restricted to permeable zones in the host rocks. The nature and geometry of such zones lead to simple and complex systems of veins, stockworks, and impregnations. Veins and vein systems are more or less tabular bodies, stockworks are rock volumes incorporating many small interconnected veinlets, and impregnations are rock volumes that have been replaced or had their pores filled by later-deposited mineral matter.

Geologists have long recognized mineral deposits formed through the action of hydrothermal solutions and described them in terms of the depth and temperature at which they formed (Fig. 3-16).

Ore minerals commonly found in hypothermal deposits (see Appendix II for a tabulation of ore minerals and their contained elements) are gold, wolframite, scheelite, pyrrhotite, pentlandite, pyrite, arsenopyrite, loellingite, chalcopyrite, sphalerite, galena, cassiterite, bismuthinite, uraninite, and cobalt and nickel arsenides. Because

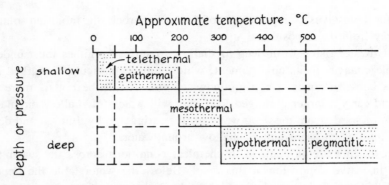

Figure 3.16. Classification of hydrothermal deposits. (*Source:* From William H. Blackburn and William H. Dennen, *Principles of Mineralogy.* Copyright © 1988 Wm. C. Brown Publishers, Dubuque, Iowa. All Rights Reserved. Reprinted by permission)

these ore minerals are deposited at considerable depths and are exposed at the surface only through deep erosion or orogenic processes, they are more common in older, often metamorphosed geologic terranes.

The Broken Hill District of New South Wales, Australia, is an example of a major hypothermal ore deposit. Since its discovery in 1883 it has produced ore worth over a billion dollars, has an annual production in excess of a million tonnes, and has a developed ore reserve of over 10 million tonnes. The sulfide ore runs about 15% lead, 12% zinc, and 5 ounces of silver per tonne. In addition to these metals, annual production also includes 60,000 tonnes of sulfuric acid, 200 tonnes of cadmium, and lesser amounts of gold, antimony, cobalt, and copper.

The deposit has the form of a saddle reef occupying the crestal zone of a tight, doubly plunging anticlinal fold in Precambrian rocks (Fig. 3-17).

The great gold deposits of the Witwatersrand of South Africa have generally been thought to be a fossil placer deposit, but recent work by Phillips and coworkers (1987) suggests hypothermal solutions of metamorphic origin have been responsible for the deposition of the gold, uranium oxides, and pyrite that characterize the ore suite.

Mesothermal deposits take many forms including disseminated ores, veins, pipes, mantos, and irregular replacement masses. The most abundant ores found are of copper, lead, zinc, silver, and gold with gangues of quartz, pyrite, and carbonate minerals. Because the ore-bearing solutions that form these deposits are typically out of equilibrium with the country rocks both chemically and thermally, extensive alteration of the wall rocks is common. The alteration products include sericite, quartz, calcite, dolomite, pyrite, orthoclase, chlorite, and clay minerals.

The variety of mesothermal deposits makes it difficult to characterize them with a single example; however, the dominance of disseminated copper deposits as a source of this metal suggests that the greatest copper deposit known, Chuquicamata, should serve. This mine is in the foothills of the Andes Mountains in northern Chile at an elevation of 2,900 meters and was known before the arrival of the Spanish conquistadores, although not seriously worked until 1879.

The ore emplacement at Chuquicamata is related to a period of faulting and ig-

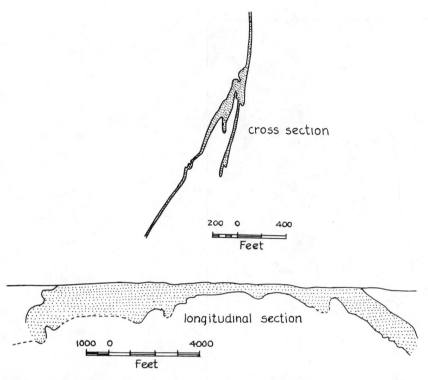

Figure 3.17. Cross and longitudinal sections (different scales) of the Broken Hill District, New South Wales. (*Source:* After B. R. Lewis, P. S. Forward, and J. B. Roberts, "Geology of the Broken Hill District," reinterpreted. *Eighth Commonwealth Mining and Metallurgical Congress,* vol. 1, 1965)

neous intrusion in late Mesozoic or early Tertiary time when several plutons, at least one of which contained small amounts of copper, were emplaced. Alternating periods of silicification and fracturing provided space for ore deposition in the Chuquicamata Porphyry and the West Fissure (Fig. 3-18) tapped copper-bearing solutions that migrated eastward into the fractured mass. Much of the primary ore at Chuquicamata has been enriched by supergene processes described later and because of the dry climate the orebody contains many rare, soluble copper sulfates, carbonates, and oxides.

Chuquicamata is but one of many disseminated copper deposits—porphyry coppers—that have formed in plutons above subducting plates where igneous rocks of intermediate (granodioritic) composition are formed from magmas generated by fractional melting of oceanic basalts (Fig. 2-2), and presumably somewhat enriched in copper. As these magmas rise, the thermal regime changes drastically, internal fluids react with earlier-formed minerals to cause extensive deuteric alteration, and groundwater circulation is greatly modified.

Should submarine volcanism occur above the intrusion, sedimentary exhalative stratiform metalliferous deposits represented by Kuroko, Besshi, and Cyprus-types of copper deposits may form on the sea floor.

Epithermal deposits are usually formed at depths less than 1,000 meters so are

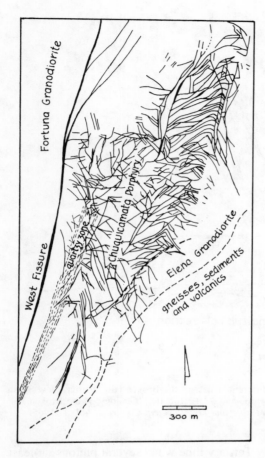

Figure 3.18. Geology of the Chuquicamata orebody. (*Source:* Redrawn from V. D. Perry, "Geology of the Cuquicamata orebody," *Mining Engineering*, vol. 4, 1952)

characteristic of regions of recent igneous activity in which erosion has not removed them. Typically, the deposits have the form of veins, irregular fissure fillings, stockworks, or breccia pipes. Often the fissure systems have a direct connection to the surface and some hot springs and steam vents, including submarine vents, are the surface expression of subsurface epithermal systems.

Ore minerals of epithermal deposits include sulfosalts of silver, tellurides of silver and gold, stibnite, acanthite, cinnabar, electrum (gold-silver alloy), and native copper. Wall rock alteration is often extensive with chlorite, sericite, alunite, zeolites, adularia, chalcedonic silica, and pyrite as characteristic products.

Epithermal silver mines at Guanajuato, Mexico, have been worked since 1548 and ore valued at more than half a billion dollars has been produced. Most of the mineralization is in a 5-kilometer stretch of the Veta Madre (mother lode), which is a vein system in a fault that strikes NW-SE, dips about 45°, and extends for 25 kilometers. Ground preparation for the emplacement of the bonanza ores occurred when the fault motion on an irregular surface generated local volumes of breccia in the hanging wall. The ores are thus distributed as stockworks in shoots a few hundred meters in area and some tens of meters thick.

The ore minerals, typical of this kind of deposit, are base metal sulfides together with silver sulfides and selenides. The gangue is pyrite, quartz, carbonates, and adularia. Both ore and gangue minerals are fine-grained and commonly occur in crustified bands.

Guanajuato is but one of the epithermal silver deposits found in the western cordillera of North America (Fig. 3-19). Similar deposits are also present in the Andes and include such famous mines as Potosi, Chocaya, and Huanchaca in Bolivia and Colquijirca in Peru.

Water in the near surface environment and far removed in space and time from any obvious connection with a magmatic source might be juvenile, meteoric, or mixed. However, as earlier stated, the distinction in the context of this book is not important and is not pursued. Several important kinds of ore deposits fall in this never-never land of low temperature, near-surface, water-formed deposits. One is the Mississippi Valley-type (MVT) lead-zinc deposit, which is characteristically formed by replacement of carbonate rocks or by open-space filling near the up-dip edges of deep basins filled with sedimentary rocks. Table 3-2 shows the stratigraphic distribution and type of ore host for the major MVT districts in the United States.

Another type of telethermal deposit is represented by the uranium-vanadium deposits such as are found on the Colorado Plateau in permeable zones in terrestrial sediments. Copper and fluorite ores of similar origin are also known.

The mineralogy of such ores is basically simple and nondiagnostic, although the number of altered varieties of uranium-vanadium minerals is huge. Wall rock alteration is limited and rock permeability is always a key to ore concentration. Figure 3-20 shows a typical uranium deposit of this kind.

Groundwater phenomena of great importance are to be found at and above the water table. At the water table there is an abrupt change in the chemical environment from reducing below to oxidizing above, countercurrent water motions occur with

Figure 3.19. Epithermal silver deposits in the U.S. and Central America.

Table 3-2.
Stratigraphic Distribution and Type of Ore Host for MVT Deposits in the U.S.[1]

Period	Producing Area	Mineralization Control
Pennsylvanian		
---unconformity---		
Mississippian	Tri-State District	Dissolution collapse breccias under unconformity
Devonian	Pine Point, Northwest Territory	Stromatolite reefs above unconformity with Precambrian
Silurian		
Upper Ordovician		
---unconformity---		
Middle Ordovician	SW Wisconsin-NW Illinois	Dissolution collapse with "pitches and flats" between unconformities
---unconformity---		
Lower Ordovician	East Tennessee; Friedensville, PA	Dissolution collapse breccia under unconformity with Middle Ordovician
Upper Cambrian	Southeast Missouri	Talus, pinchouts, reefs, and sedimentary breccias above unconformity with Precambrian
Lower Cambrian	Austenville, VA	Reefs and breccias at major facies changes extending up through the section. No unconformities identified
---unconformity---		
Precambrian		

[1] Personal communication, Peter E. Price, 1985.

downward infiltration and upward capillarity, and the physical system is tripartite—solid, liquid, and gas. The phreatic zone below the Earth's surface and above the water table is one of great chemical activity where essentially all of rock weathering occurs. These processes have been earlier described for ordinary rocks, but some remarks on the reaction of mineralized deposits in this zone should be made. Of particular interest is the response of rocks containing sulfide minerals to conditions in the phreatic zone. Many metallic sulfides oxidize readily in an environment containing oxygen and water, that is, the sulfide, S^{2-}, compound reforms as a sulfate, SO_4^{2-}, or oxygen, O^{2-}, compound. These oxides and sulfates may have different solubilities so insoluble species may remain where formed, whereas soluble matter migrates downward and outward with the groundwater under the influence of gravity and capillarity.

For example, consider the series of reactions for chalcopyrite, an important primary ore of copper, in the zone of weathering.

1. Chalcopyrite, $CuFeS_2$, in the presence of water and oxygen weathers to yield insoluble red iron oxide plus copper and sulfate ions in solution:

$$2CuFeS_2 + 9O_2 + H_2O = Fe_2O_3 + 4SO_4^{2-} + 2H^+ + 2Cu^{2+}$$

Groundwater and Ore Solutions

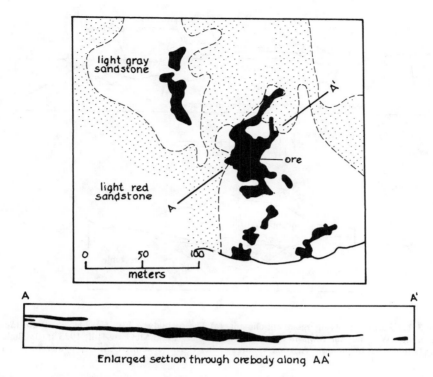

Enlarged section through orebody along AA'

Figure 3.20. Stratabound uranium deposit. (*Source:* Redrawn after P. H. Dodd, "Examples of uranium deposits in the upper Jurassic Morrison Formation of the Colorado Plateau," *U.S. Geological Survey Professional Paper* 300, 1956)

2. The iron oxide remains in place giving a characteristic red color and cellular texture to the leached zone or *gossan* (old Cornish usage), whereas copper and sulfate ions migrate downward. At the water table they encounter a reducing environment and recombine to form a blanketlike deposit of a new solid, chalcocite, Cu_2S, which is free of iron and contains proportionately more copper than the original chalcopyrite:

$$2Cu^{2+} + SO_4^{2-} + 8H^+ = Cu_2S + 4H_2O$$

The presence of a gossan at the surface thus marks the subsurface position of intense chemical activity in which separation and concentration of different elements is accomplished by their different solubilities. Some elements such as copper are concentrated downward to the water table; others form insoluble oxidized compounds, for example, silver chloride or lead sulfate, and are enriched in the phreatic zone by the leaching out of other substances. Figure 3-21 shows cross-sections of two major copper deposits made workable by supergene enrichment of subeconomic mineralization.

It must always be remembered that because of the intense chemical activity in the

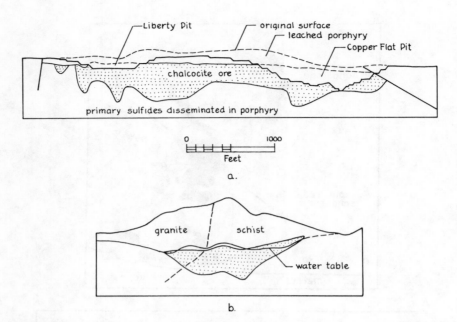

Figure 3.21. Examples of supergene enrichment. (a) The orebody at Ely, Nevada, ca. 1920. (*Source:* Redrawn after W. Lindgren, *Mineral Deposits*, 4th ed. New York: McGraw-Hill, 1933.); (b) Chalcocite blanket (stippled) at Inspiration, Arizona. (*Source:* After F. L. Ransome, *U.S. Geological Survey*, 1919)

oxidizing zone, the mineral matter there may show little if any similarity to the underlying bedrock or mineral deposit. Further, the water table is not fixed in relation to the surface but will rise or fall depending on climatic and tectonic changes.

DEPOSITION OF THE ORES

Elements that have been selectively partitioned into the hydrous phase under some particular set of conditions will remain in solution only so long as those conditions obtain. As the ore-bearing water moves, it must encounter new environments, and reactions to maintain its chemical equilibrium with its host rock will take place. The process is analogous to the lithification of sedimentary rocks by cementation. The particular environmental conditions of importance in ore deposition will differ from one to another location and ore, but will probably be one or a combination of the following:

    Chemical causes of deposition:
        changes in pH
        changes in Eh
        reaction with solids
        reaction with other liquids

Physical causes of deposition:
  drop in temperature
  change in confining pressure
  dilution by unsaturated water
Biologic causes of deposition:
  bacterial activity

Calculations and experiment indicate that pH, the activity of hydrogen ions and commonly called acidity, together with Eh, the oxidizing or reducing tendency of a solution, are of considerable importance in ore deposition. Changes in these parameters might be brought about, for example, by chemical reactions that alter the concentration of hydrogen ions, say by reaction of the H^+-bearing solution with calcite

$$H^+ + Ca(CO_3) = HCO_3^- + Ca^{2+}$$

or alter the oxidation state, say for iron from the ferrous to ferric state

$$4Fe^{2+} + 3O_2 = 2Fe_2O_3$$

The ore solution may react with the rock material through which it is passing or encounter chemically different solutions. A favored host for solution-rock reactions is a carbonate rock that is sometimes completely replaced by metal sulfides. Solution-solution interaction may be particularly important when saline waters of deep circulation encounter the fresher, cooler, and possibly sulfate-bearing groundwater of shallow circulation. In the latter instance, reactions with sulfate ions to form sulfide and sulfate ore minerals will occur, for example,

$$Ba^{2+} + SO_4^{2-} = Ba(SO_4) \quad \text{(barite)}$$

or

$$Zn^{2+} + SO_4^{2-} = O_2 + ZnS \quad \text{(sphalerite)}$$

The oxidation state of many elements, for example, arsenic, chromium, copper, iron, manganese, mercury, sulfur, vanadium, and uranium, is readily changed under geologically reasonable conditions. Such changes always involve a difference in the solubility of the compounds so oxidation-reduction (redox) reactions can be very effective in ore deposition. An example would be reduction of uranium in a soluble U^{6+} complex to an insoluble quadrivalent form in uraninite, UO_2, by the action of carbonaceous material.

Physical changes may be instrumental in the deposition of ore minerals from solution. The solubility of most substances increases with increasing solvent temperature so precipitation occurs as an ore solution moves upward and cools. Prediction of depth as a function of temperature is, however, difficult because water is an excellent carrier of heat and can easily bring temperatures in a conduit to levels well above those of the wall if introduced rapidly. Further complications are introduced by the heat-evolving (exothermic) nature of most reactions involving the crystallization of ore and gangue minerals.

The actual temperature at which a mineral was formed may be ascertained in a number of ways, of which the study of fluid inclusions is the most widely used. Minerals may surround and entrap foreign matter as they crystallize and may be seen to be inclusion-charged under a microscope. Among the included matter may be a liquid, which, if no means for its access is present, must be a sample of the ore-forming solution. Such fluids are typically saline and lead to the identification of salty water with ore solutions.

Liquids show much greater changes in volume with temperature than do solids. Hence, as the host mineral cooled, the trapped liquid shrank more than the vacuole in which it was trapped. To find the temperature of a mineral's formation, the fluid inclusion is observed as the mineral is slowly heated. The temperature at which the expanding liquid fills the vacuole is the temperature at which the host mineral crystallized.

Changes in pressure usually have less effect on ore deposition than do changes in temperature. Pressure differences can be effective, however, under such special circumstances as solubility control by gases such as H_2S or CO_2 dissolved in water or abrupt adiabatic pressure changes as the fluid moves through a constriction.

Deposition of ore and gangue minerals from ore solutions may not only be the result of physical changes and inorganic chemical reactions, but also takes place by biochemical means. In many and a growing number of instances bacteria have been shown to play a major role in the formation of particular mineral deposits.

The classic identification of bacteria with ore genesis is their role in depositing native sulfur in the cap rocks of salt domes. Deeply buried salt beds deform plastically under the weight of superincumbent rock and penetrate upward as pipelike plugs called salt domes, bending upward the layers they penetrate. Some dissolution of the rock salt at the top of the dome occurs leaving less soluble anhydrite, $Ca(SO_4)$, which is present in the salt mass, as a capping.

The emplacement of a salt dome provides an ideal circumstance for the migration and entrapment of fluids such as oil and natural gas, which rise in an annular ring around the impermeable dome of salt and are trapped under aquacludes in the upturned beds or in reservoirs formed by the extensive faulting associated with the emplacement of the salt dome.

Anaerobic sulfate-reducing bacteria (*Desulfovibrio desulfuricans* and *Clostrium nigrificans*) are found in a wide variety of habitats where their presence is marked by black coloration and a smell of sulfide. They reduce sulfate ions to hydrogen sulfide gas using organic material as an energy source. In the cap rocks of salt domes these bacteria actively reduce the sulfate ions from anhydrite and generate hydrogen sulfide gas, H_2S, using upward-seeping petroleum fluids for energy.

Chemautotrophs, bacteria requiring no preformed organic compounds for growth, assimilate H_2S and oxidize it to native sulfur, which can accumulate to orebody dimensions. Many such bodies exist along the Gulf Coast of the United States and Mexico as well as in Germany, Poland, Rumania, Russia, and Iran.

The presence of an impermeable salt plug not only controls the upward flow of petroleum fluids but also of groundwater. This may be metal-rich and, in the near surface zone, metals rather than chemautotrophs may react with the hydrogen sulfide to form metal sulfide. The orebodies on the Hockley and Winfield Domes in Louisiana have this origin.

Many orebodies that occur as thin extensive horizons in sedimentary rock sequences are known or suspected as being of bacteriogenic origin. Generally the rocks contain carbonaceous material suggesting a depositional environment in which bacteria could flourish; the source of the metal-rich solutions is, however, often controversial. Examples of these deposits include the world-class base metal sulfide deposits of Mount Isa in Australia and White Pine, Michigan; the cupriferous sedimentary rocks of the Zambian Copperbelt and the Kupferschiefer of Germany, as well as numerous red-bed copper deposits; uranium concentrations in sandstones and black bituminous shales (Chattanooga Shale of Tennessee and adjacent states, Kolm of Sweden); and deposits of both iron sulfide (Nairne, South Australia), iron oxide, and manganese minerals.

Major deposits for which bacterial action is thought to be involved in the precipitation of base metal sulfides on the sea floor from volcanic exhalations or waters driven out from sedimentary piles undergoing lithification include Kidd Creek, Ontario, Sullivan, British Columbia, Buchans, Newfoundland, and Flin Flon, Saskatchewan-Manitoba in Canada, Skellefte, Sweden, and the Kuroko deposits of Japan.

A traditional practice in copper mining areas has been the recovery of copper metal from various copper-rich drainage waters by its exchange with iron. Copper ions in solution will replace iron atoms in a solid so scrap iron is gradually transformed to copper when in contact with copper-bearing waters. Much of the drainage water in mining activities comes from waste dumps in which subgrade copper minerals in the broken rock can be readily attacked by air and water. Modern practice in large operations, for example, Bingham Canyon, Utah, is to accelerate the leaching of the dumps by judicious seeding with bacteria.

References

Baskov, E. A. 1987. *The Fundamentals of Paleogeology of Ore Deposits.* New York: Springer-Verlag.

Barnes, H. L., ed. 1979. *Geochemistry of Hydrothermal Ore Deposits*, 2nd ed. New York: John Wiley & Sons.

Bickle, M. J., and MacKenzie, D. 1987. The transport of heat and fluids during metamorphism. *Contributions to Mineralogy and Petrology*, 95:384–392.

Dennen, W. H., and Moore, B. R. 1986. *Geology and Engineering*. Dubuque: Wm. C. Brown.

Ferry, J. M. 1980. A case study of the amount and distribution of fluid during metamorphism. *Contributions to Mineralogy and Petrology*, 71:373–385.

Freeze, R. A., and Cherry, J. A. 1979. *Groundwater*. Englewood Cliffs, NJ: Prentice-Hall.

Fyfe, W. S., Price, N. J., and Thompson, A. B. 1978. *Fluids in the Earth's Crust*. Amsterdam: Elsevier.

Goldfarb, R. J., Pickthorn, W. J., and Paterson, C. J., 1988. Origin of lode-gold deposits of the Juneau gold belt, southeastern Alaska. *Geology*, 16:440–443.

Goguel, J. 1976. *Geothermics*. New York: McGraw-Hill.

Koslovsky, Ye. A., ed. 1987. *The Superdeep Well of the Kola Peninsula*. New York: Springer-Verlag.

Phillips, G. N., Myers, R. E., and Palmer, J. A. 1987. Problems with the placer model for Witwatersrand gold. *Geology*, vol. 15.

Sibson, R. H., McMoore, J., and Rankin, R. H. 1975. Seismic pumping—a hydrothermal fluid transport mechanism. *Jour. Geol. Soc. London*, 31:653–659.

Chapter 4
Planning, Exploration, and Geologic Evaluation

If you're going to prospect you've got to have intelligence,
BUT
if you've got intelligence you don't have to prospect!
Tiger Olsen

INTRODUCTION

Mineral exploration is a continuing activity of individuals and specialized groups who are attempting to find mineral resources in advance of need and also to develop new or improved methods of discovery, mineral extraction, beneficiation, or transport. The level of such activity may be more or less intense as a function of market demands, but the long lead time, typically 10 to 15 years, between discovery and production requires ongoing exploration programs.

An often employed rule of thumb is that 5% of the gross anticipated value of a deposit may reasonably be spent on its discovery. Seldom, however, is such a sum budgeted at the outset; usually money for initiation is allocated and additional outlays made following favorable assessment of the information acquired. Exploration programs must thus be continuously justified and may be terminated at any stage. This feature is typically not understood by landholders on whose property investigations are made—that is, work must be performed to establish value, but a study cannot add wealth and a potentially unprofitable venture should not be pursued.

Individual or corporate decisions to follow a particular line of exploration may be based either on the present or anticipated need for a particular commodity or on the desire to obtain an evaluation of the mineral potential of a particular area. In the former instance, targeting takes no initial account of geographic or political boundaries and may, indeed, be initiated on a worldwide scale. Alternately, it may be desired to know of all potential resources in some geographically defined area. Obviously, these are not mutually exclusive approaches, but they do serve to identify the principal bases for selection of the exploration approach and methods to be used.

The goal of mineral exploration is to reduce large areas to testable targets. Ore deposits may be quite small—a million dollars in gold will fit into an ordinary wastebasket—so the process is comparable to finding the proverbial needle in the haystack. Fortunately, very sensitive exploration tools are available and may be distinguished as geologic, geochemical, geophysical, or remote sensing techniques. Typically two or more are employed in company or sequentially during any exploration program.

EXPLORATION GEOLOGY

Mineral deposits are by their nature different in some degree from other rocks in their mineral content, and this difference, in turn, gives the deposit distinguishable chemical and physical properties. Exploration for a deposit may thus be directed toward the location of areas of earth materials showing nonaverage or anomalous properties. Some of the principal tools used in this search are briefly described later in this chapter. In a general way the procedure is to first identify areas showing deviation from the general physical or chemical properties of the rocks or soils of the survey region and then evaluate the cause of the anomaly. Since rock properties are inherently variable and the works of humans can confuse and contaminate the data, it is usual experience that many meaningless anomalies are discovered in a survey, such geologically nonuseful information as the location of dumps, culverts, fences, or graveyards, for example. Undoubtedly the most difficult part of an exploration program is not the discovery of anomalies but of assessing their significance.

Geologists serve as exploration theorists in the search for ore and usually use either a genetic or a comparative model in their approach. Obviously, an understanding of the development of rock sequences in time coupled with recognizable rock-ore associations should lead to an identification of the more probable locations and kinds of ore to be found in a particular region. There is, for example, little point in looking for tin mineralization outside a region of salic igneous rocks or hydrocarbons in a high-grade metamorphic terrane.

It is a truism in mineral exploration that one should seek elephants in elephant country, that is, the likelihood of another deposit being found in a known mineralized district is greater that an equivalent discovery in a virgin area. The continued viability of some mining districts has been truly remarkable with initiation, abandonment, and rebirth of mining activity sporadically occurring over centuries following new discoveries, different commodity requirements, or changing economic conditions. On the other hand, such world-class deposits as Carlin, Nevada (gold), Weipa (bauxite), and Olympic Dam (copper-uranium-gold) in Australia, Hemlo (gold) and Kidd Creek (base metals and silver) in Canada, and Trombetas (bauxite) and Carajas (iron ore) in Brazil have been found in "rabbit country" in recent years.

If an ore deposit is already known, its nature and geologic controls may be used as a reference against which other observations in the area are compared. For example, base metal mineralization in eastern Maine is controlled by minor faults or folds near the boundaries of a particular type of granite where it is intrusive into magnesia-rich rocks. The assemblage of general geologic features that correlate with mineralization are termed *ore guides* and their recognition is an essential step in exploration using the comparative approach.

During the course of an exploration program, the geologist must acquire an intimate knowledge of the target area from a study of all of the pertinent literature and by doing fieldwork. From this background a preliminary interpretation is made that serves to direct the continuing geologic work and that of the various supporting specialists. Reinterpretation will follow acquisition of new data that weakens or strengthens a previously held concept, and an initial broad area with multiple possibilities is gradually reduced to a small number of localities worthy of detailed on-site study.

Prospecting methods fall naturally into those employing remote sensing, geologic studies, and geochemical and geophysical methods. These approaches are applicable to the discovery of any mineral deposit, but, for simplicity, the following discussions will generally assume the target to be a metallic ore deposit.

On-site examination is still the cheapest but probably the slowest method of geologic exploration. Fieldwork to discover suggestive rock types or structures or to find indications of mineralization is sometimes used as the first exploration step, but more often follows more sophisticated techniques. In most areas of the world field geologists have maps and aerial photographs available to guide their work, and with them the regional geology may be assessed on a general basis and a reasoned approach to fieldwork planned.

Geologic mapping usually begins with reconnaissance traverses and sampling to locate favorable rocks, rock associations, and structures and is followed by detailed mapping when they are located. The main tools are a compass and clinometer, notebook, hand lens, hammer, and sample bags by means of which rocks are sampled and labeled, and observations made and recorded. In some areas mapping may produce relatively quick results; in others of complex geology or hostile climate, it can take years.

Until recently, many mining companies, large and small, hired prospectors to hunt for suitable rocks and indications of mineralization. They were usually miners or partly trained geologists who searched large areas or existing prospects in close detail as did the prospectors at the turn of the century. Their results were then interpreted by company geologists who took to the field when promising ground was located. Most of the major mines of the world found prior to the 1960s were discovered by following up indications given by prospectors.

GEOPHYSICAL EXPLORATION METHODS

Mineral deposits by their special nature often have physical properties that are sufficiently distinctive to be used for their discovery and delineation. The properties that have been found to be particularly useful in mineral exploration are radioactivity, magnetic susceptibility, seismic (sound) velocity, and electrical resistance, conductance, and potential difference. The principal geophysical prospecting methods are thus divisible into radiometric, seismic, magnetic, electrical, and electromagnetic methods (Table 4-1). Geophysical observations can be made on the ground, in the air, or at sea, and although the equipment varies in the three different situations, the geophysical principles remain the same.

In prospecting, the physical property or field of interest is sampled in some regular way in order to discover any irregularities or anomalies that may be ascribed to the hidden presence of a deposit of interest. Some generalities of importance are that certain properties such as the Earth's magnetic intensity may not be constant in time, gravimetric data requires corrections for the mass of nearby topography, and electrical properties are greatly affected by the water content of rocks. As a rule, geophysical methods involve the sampling of a field whose intensity decreases as an inverse square and so decays rapidly with depth to a buried body and with the height of airborne equipment above ground.

Magnetic methods depend on the fact that different rocks and, in particular, some buried orebodies, locally affect the Earth's magnetic field and produce an anomaly of either a positive or negative kind, thus indicating a possible exploration target. The most magnetically distinct anomalies are due to the presence of iron ores that

Table 4-1.
Summary of the Principal Geophysical Methods

Method	Targets
Gravitational	Structures (folds, buried ridges, faults), intrusions
Magnetic[1]	Structures, magnetic and associated ores (sulfide orebodies, placers)
Electrical	
Resistivity	Sulfide orebodies, water table, subsurface profiles
Self-potential (SP)	Sulfide orebodies
Induced potential (IP)	Disseminated ores, sulfide orebodies, groundwater, structure
Electromagnetic (EM)[1]	
Natural field	
Audiofrequency magnetic field (AFMAG)	Profiling, highly conductive ore bodies
Very low frequency (VLF)	Profiling, structure, conductive orebodies, contamination plumes
Induced field	
Compensator and Turam	Electrically conductive orebodies
Seismic	
Reflection	Structure, profiling
Refraction	Structure, profiling
Radiometric[1]	Radioactive ores and associated materials, regional discontinuities
Thermal	Sulfide ores, fissures, intrusions, geothermal resources

[1] May be airborne.

contain magnetite or other iron minerals with a degree of permanent (remanent) magnetism (Fig. 4-1). Many metallic orebodies also contain slightly magnetic minerals or weather to yield oxides of iron, which are detectable magnetically in a magnetometer traverse.

The simplest magnetometer traverse is conducted using a hand-held or tripod-mounted magnetometer that can easily be carried. Care is taken not to take meter readings near surface magnetic features such as metallic structures, fences, power lines, or vehicles or to carry magnetic objects. An airborne rig with various types of sensors trailed behind the aircraft as drogues to obviate magnetic interference from the plane, and recording equipment mounted in the aircraft is much more expensive but can cover vast areas of ground rapidly. Besides the geophysical instruments, airborne units require complex navigation and data reduction systems. Similar recording equipment can be mounted in ships and detection equipment trailed behind by cable for offshore work and, in this case, seismic observations can also be made that are not possible from an aircraft.

The Earth's magnetic field shows both local and regional variation in its orientation and strength in time as a natural consequence of its interaction with materials of different permeability and with the Earth's internal and external electrical fields.

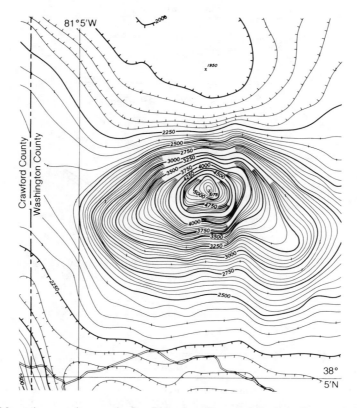

Figure 4.1. Magnetic anomaly over the Pea Ridge Iron Deposit, Missouri. (*Source: U.S. Geological Survey*)

Magnetic anomalies of prospecting interest are superimposed on this irregular and slowly varying background. Because the field lines are inclined to the surface except at the equator, magnetic measurements should distinguish between vertical, horizontal, and total intensity as well as polarity and direction.

Seismic prospecting is used intensively in the petroleum industry for detecting buried geologic structures that may contain oil or gas and is also used for the interpretation of rock structure in mineral exploration. The principle is that artificial shock waves are induced in the ground by explosives or a hammer blow and detected by strategically placed receivers called geophones. Analysis of travel times of reflected or refracted shock waves allows an interpretation of the depth to boundaries of rock units having contrasting seismic velocities and of the configuration of the underlying rocks (Fig. 4-2). The same principle is employed in offshore exploration with one ship dropping the explosive charges and another running parallel towing a recording cable connected to equipment on board. Almost all of the major offshore oil fields have been found in this way.

The gravitational pull on an object at or near the surface of the Earth varies as a

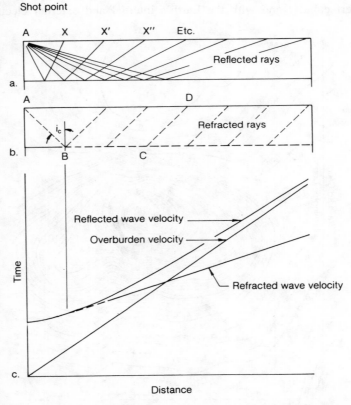

Figure 4.2. Travel paths and times of seismic waves (velocity in lower layer twice that of Upper) Key: ABCD = refracted wave path. AX, AX', AX'' = reflected ray paths. i_c = critical angle. (*Source:* From William H. Dennen and Bruce Moore, *Geology and Engineering*. Copyright © 1986 Wm. C. Brown Publishers, Dubuque, Iowa. All rights reserved. Reprinted by permission)

function of the density of the rocks in the crust at a particular point. Very sensitive instruments called gravimeters are used to measure variations in the gravity field from point to point and the resultant map may be interpreted in terms of underlying rock types and structure. Careful corrections must be made to the measured data because of the differences in field strength as a function of distance from the Earth's center (latitude and altitude) and deficiencies or increments of mass due to irregular terrain.

There are a number of electrical properties of earth materials that are readily measurable and can serve as exploration tools. The geophysical techniques that have been developed may be divided into passive methods in which the properties of the ground are directly measured (e.g., self-potential) and active methods where the effects of a stimulating electrical field, applied by direct contact or inductance are measured (e.g. induced potential, resistivity, electromagnetic methods).

The apparatus used for resistivity measurements consists of batteries or hand-cranked generator, a double commutator, four electrodes, a milliammeter, and a potentiometer, or a direct reading ohmmeter. Current through the milliammeter, current electrodes, and ground is periodically reversed to obviate polarization at the electrodes and the current through the potential (sampling) electrodes is simultaneously reversed so d.c. potentials are read on the potentiometer. The electrode configuration used for mapping (Fig. 4-3) is moved along a traverse line and provides constant depth penetration. Increasing the separation of the electrodes provides increasing penetration and, if symmetrically done, allows "resistivity drilling" to be performed. The electrical relations are given by

$$\rho = 2\pi\, aV/I = 2\pi\, aR$$

where ρ = resistivity of hemisphere of radius a = length of current path = electrode spacing, V = voltage, I = current, and R = resistance.

An example of resistivity mapping is given in Figure 4-4 where the presence of anhydrite in a gypsum body and the thickness of clay overburden has been deter-

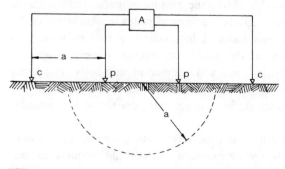

KEY

A   Generator, commutator, meters
c   Current electrodes
p   Potential electrodes
a   Approximate depth of effective measurement

Figure 4.3. Arrangement of components for resistivity mapping. (*Source:* From William H. Dennen and Bruce Moore, *Geology and Engineering.* Copyright © 1986 Wm. C. Brown Publishers, Dubuque, Iowa. All rights reserved. Reprinted by permission)

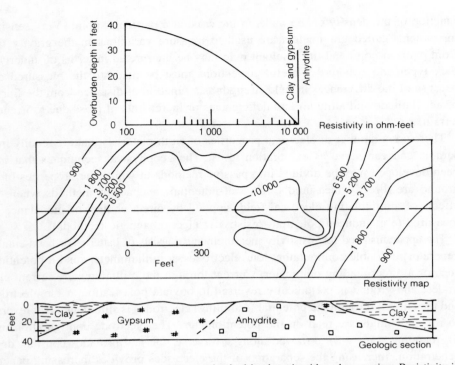

Figure 4.4. Resistivity contours on a gypsum and anhydrite deposit with a clay capping. Resistivity in ohm feet: >10000—anhydrite. 6500–10000—gypsum, 2 feet of overburden. 5200–6500—gypsum, 5 feet of overburden. 3700–5200—gypsum, 10 feet of overburden. 1800–3700—gypsum, 20 feet of overburden. (*Source:* From William H. Dennen and Bruce Moore, *Geology and Engineering.* Copyright © 1986 Wm. C. Brown Publishers, Dubuque, Iowa. All rights reserved. Reprinted by permission)

mined. Measured and contoured resistivity values have been independently correlated with rock type and overburden thickness.

Electromagnetic (EM) methods are designed to measure the distortion of primary electromagnetic fields caused by induction of electric currents in buried conducting bodies. The primary field may be introduced into the ground by natural atmospheric disturbances (worldwide thunderstorm activity), powerful radio transmitters, or by induction coils. The alternating currents induced in conductive materials such as graphitic zones or sulfide bodies become the source of secondary electromagnetic waves that can be detected by suitably designed and oriented pickup coils. The depth of penetration of electromagnetic waves into the earth depends on the resistivity of the ground and frequency of the radiation, increasing with resistivity and lowered frequency.

A large number of EM methods are in use because both airborne and ground systems can be deployed, there exists a wide range of sources and frequencies that can be employed, and different kinds of measurements can be made in the field. Some methods employ fixed sources and moving receivers and in others both source and receiver are moved in some regular way.

The VLF (very low frequency) method has enjoyed considerable popularity in recent years and may be taken as an example of an EM prospecting tool. The method uses carrier wave signals broadcast for marine and air navigation systems as the source of its primary field. These signals are in the range of 15–25 kHz and can be detected at useful levels for distances of thousands of kilometers. Suitable transmitting stations for EM prospecting in North America are located in Hawaii; and in Seattle, Washington; Annapolis, Maryland; and Cutler, Maine.

A VLF receiver employs two coils (X and Z) fixed at right angles in the form a cross. In use, the instrument is oriented by setting the Z coil, normally held vertically, in a horizontal position and rotating it in the horizontal plane to find the direction to the transmitter, which is indicated by a minimum signal. Next, the Z coil is held vertically, X oriented along the line of transmission, and Z tilted about the axis defined by X. The signal from the X coil is electronically shifted in phase, usually by 45° or 90°, in the receiver and connected in series with the signal from the Z coil. The amplitude of the Z signal is adjustable and, by its adjustment coupled with measurements of tilt made with a built-in clinometer, a minimum signal and tilt angle is found. Moving the receiver along traverse lines perpendicular to the transmission direction and preferably normal to the strike of the conducting body allows its boundaries to be defined.

GEOCHEMICAL PROSPECTING

Concentrations of valuable substances in ore deposits imply an anomalous concentration of one or more chemical elements. The actual level may be as small as a part per million for some rare elements of value or be a significant fraction of the total rock. Ore deposits may be relatively small and hidden under surface cover. However, the deposit will typically be subjected to weathering and erosion and the resulting dispersion of its constituents greatly enlarges the target. Prospecting by following glacial erratics or stream-transported boulders upstream to their source has been employed for hundreds of years and the use of panning as a guide to locate deposits of gold, tin ore, or diamonds is a much-used technique. In recent years, the identification of dispersed constituents by the use of sensitive and selective chemical testing of soil, stream sediments, and vegetation has been employed with considerable success.

A metallic deposit in the zone of weathering releases ions that are dissolved and transported in the ground and surface waters. Exchange reactions, principally adsorption, build up low level "contamination" in the soil and on stream-borne particles in the vicinity of the ore deposit source, and this anomalous metallic content may be found by appropriate testing. Experience has shown that silt is the best material for stream sampling and that the most useful soil samples are obtained from the B horizon. Unfortunately, not only ore deposits but also the works and activities of humans release metals into the environment. The geochemical sampler must always be alert to the presence of such things as fertilized fields, cemeteries, dumps, culverts, waste outfalls, and fish hatcheries.

Direct sampling of water has been used on occasion, but, except in special circumstances, the variation in flow of streams causes too much variation in the concentration levels of dissolved metals and the distribution of lakes, wells, and springs is too erratic. Direct sampling of vegetation has been used for many years but with limited success due to procedural difficulties, that is, samples must be of the same part of the same plant species and ashed before analysis. The subtle response of foliage color to a plant's uptake of trace elements, however, is proving of great value in prospecting by remote sensing methods.

Many analytical techniques are employed in geochemical prospecting, the more usual being atomic absorption and emission spectrography, x-ray fluorescence analysis, and various colorimetric tests that employ selective metallo-organic reactions capable of detecting the presence of low levels of adsorbed metallic ions. One such test, the popular dithizone method (cold extractable heavy metals or CXHM test) for testing the total adsorbed heavy metal content (copper + lead + zinc + cobalt) of a sample, is described below:

1. Place 0.1 gram of dried and sieved to -12 mesh or finer sample in a test tube.
2. Add 3 milliliters of demineralized 25% aqueous solution of 1:5 hydroxylamine hydrochloride : ammonium citrate, pH 8.5.
3. Add 1 milliliter of a 0.001% solution of diphenylthiocarbazone (dithizone) in organic solvent—benzene or toluene.
4. Shake 30 seconds, allow solutions to separate, and observe color of organic layer that rises to the top. If green, record 0, blue-green 1/2, and blue 1.
5. If color is purple or red, add dithizone solution with shaking until the organic layer is sky blue. Record the number of milliliters of dithizone solution added.

The sequence of events that occurs in this test are the stripping of adsorbed ions from particle surfaces by the citrate solution followed by the chelation of the heavy metal ions by the dithizone. The dithizone colored by chelated ions rises to form an organic layer, and the amount of chelated metal is determined by dilution. Preparation of solutions and applications of this and other geochemical tests are described in detail in the *U.S. Geological Survey Bulletin 1152*.

It should be noted that the testing for one metal may well lead to discovery of a deposit of another since most sulfide deposits are polymetallic. Zinc, mercury, and arsenic have proven to be particularly useful pathfinder elements, and, for example, prospecting for gold is often conducted by geochemical testing for arsenic, which commonly accompanies it.

Geochemical test results are plotted on maps and the data examined to discover chemical anomalies that may be related to ore. The recognition of significant anomalies will usually require a statistical examination of a large amount of data in order to distinguish the anomalous data from the chemical background. This is usually done by constructing a histogram or distribution curve from all of the data taken in the sampling area that has a logarithmic absissa in recognition of the lognormal distribution of the chemical elements in nature. Figure 4-5 shows such a distribution

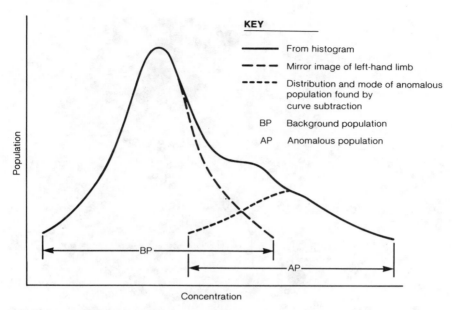

Figure 4.5. Separation of background and anomalous populations. (*Source:* From William H. Dennen and Bruce Moore, *Geology and Engineering.* Copyright © 1986 Wm. C. Brown Publishers, Dubuque, Iowa. All rights reserved. Reprinted by permission)

curve and a graphical solution to separate the background and anomalous populations.

Geochemical prospecting is employed both in reconnaissance studies of large areas, often in conjunction with the collection of geologic or geophysical data, and in detailed site examinations. Reconnaissance typically employs stream sediments as sample materials, whereas detailed studies are usually made by collecting soil samples on a regular grid. Figure 4-6 is a portion of Mineral Investigations Field Studies Map MF-301, U.S. Geological Survey, 1967, showing the results of reconnaissance sampling for heavy metals in stream sediments in eastern Maine. The size of circles indicates the amount of metal found by dithizone testing of each sampling site and the clustering of anomalous values suggests locales for more detailed exploration.

Figure 4-7 shows an idealized geochemical approach to ore finding where an area of interest is located by stream sampling reconnaissance and the ore body located by detailed soil sampling.

REMOTE SENSING

Remote sensing as the term is presently employed may be defined as the use of electromagnetic radiation for the collection of information about an object without having physical contact with it. Aerial photography is an early form of remote sensing, and recent developments have added the ability to use radiation outside the visible portion of the spectrum, employ electro-optical scanners in place of film, and

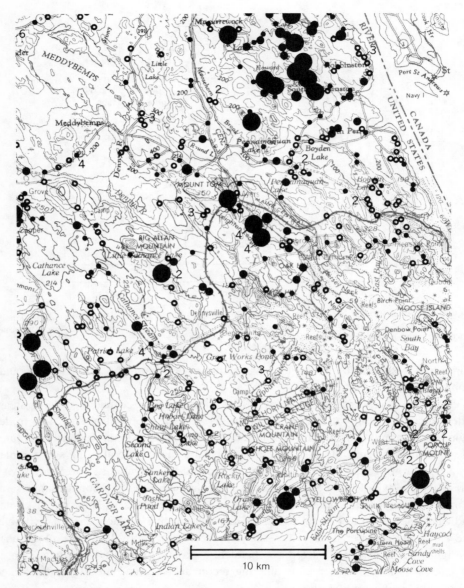

Figure 4.6. Geochemical reconnaisance map, southeastern Maine. Key: Amounts of heavy metals (Cu + Pb + Zn + Co) extracted from stream sediments. Large circles > 20 parts per million. Numbers are parts per million copper. (*Source:* From E. V. Post, et. al., Mineral Investigations Field Studies Map Mf-301, *U.S. Geological Survey,* 1967)

to apply computer processing for the formation and interpretation of images. These are new and powerful techniques in the arsenal of the ore finder. Particularly notable is the routine coverage of very large areas through the use of Earth satellites as "photographic" platforms and active systems at radar frequencies that are unaffected by weather conditions.

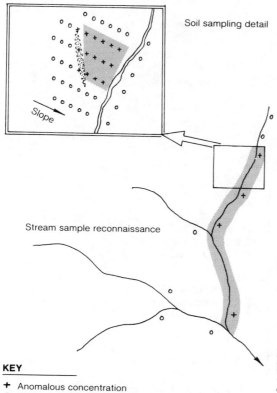

Figure 4.7. Idealized geochemical approach to ore finding. (*Source:* From William H. Dennen and Bruce Moore, *Geology and Engineering.* Copyright © 1986 Wm. C. Brown Publishers, Dubuque, Iowa. All rights reserved. Reprinted by permission)

Electromagnetic radiation moves at the speed of light with a harmonic wave motion, and electromagnetic waves of all wavelengths travel with the same velocity, nearly 300,000 km/s in a vacuum. The relationship is:

$$C = \lambda \nu$$

where C = velocity of electromagnetic radiation, a universal constant, λ = wavelength, commonly expressed in nanometers (1 nm = 10^{-9} m), micrometers (1 μm = 10^{-6} m), centimeters, or meters, and ν = frequency in cycles per second termed hertz, Hz.

Electromagnetic waves excompass a very extensive spectrum of wavelengths ranging about 10^{-12} m (gamma rays) to 300 m (radio waves). That portion utilized by one or another remote sensing technique is shown in Figure 4-8.

The interaction of electromagnetic radiation with matter as observed in its selective absorption and reflectance is quite varied and allows different information to be obtained through the use of different wavelengths. The principal limitation is in the relatively narrow bands of incident wavelength that are available in a practical way, namely solar radiation and radar waves.

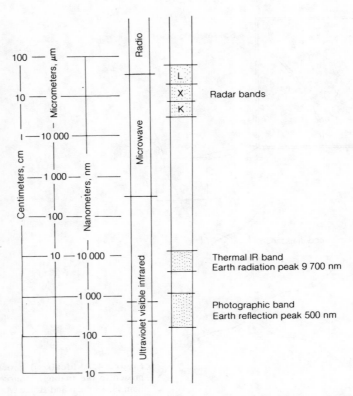

Figure 4.8. Wavelengths utilized in remote sensing. (*Source:* From William H. Dennen and Bruce Moore, *Geology and Engineering*. Copyright © 1986 Wm. C. Brown Publishers, Dubuque, Iowa. All rights reserved. Reprinted by permission)

Visible radiation and portions of the ultraviolet and near infrared spectrum, about 0.25 to 1.2 μm, may be detected by photographic means by use of different film emulsions. The most commonly used photographs are ordinary black and white; infrared black and white, which shows reflected solar radiation; infrared color, once called camouflage detection film; and ultraviolet. The latter has poor resolution because of light scattering by atmospheric dust and gas molecules but sometimes shows carbonate rocks as especially bright because of ultraviolet-stimulated fluorescence.

Aerial photography has been widely used for many purposes from map making to crop inventories and provides the geologist or engineer with detailed views at different scales of features of interest. Among its more useful aspects is the use of photo pairs or sequences to provide a three-dimensional image that greatly assists in identifying the boundaries of geologic formations, locating faults and other linear features, and showing topographic relief.

Human eyes are coupled in such a way that near objects appear three-dimensional because each eye sees a slightly different view and the brain combines them to an object at an appropriate distance (Fig. 4-9a). This stereoscopic vision may be imitated if two sequential photographs of the same object are taken from an aircraft (Fig. 4-9b). When these paired photos are viewed by eye or in a stereoscopic viewer,

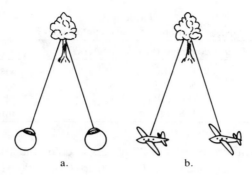

Figure 4.9. The geometry of stereoscopic vision: (a) Different views of an object as received by two eyes are blended into a three-dimensional image by the brain. (b) Different views of an object acquired photographically, perhaps from an airplane, can also be blended into a three-dimensional image, usually through the use of a stereoscope. (*Source:* From William H. Dennen and Bruce Moore, *Geology and Engineering*. Copyright © 1986 Wm. C. Brown Publishers, Dubuque, Iowa. All rights reserved. Reprinted by permission)

they are blended to a single three-dimensional scene at the apparent distance of the original. The system may be quantified if the separation between the eyes or photographic locations and the convergence angle of the viewed angle is known. This is the principle employed in optical range finders.

Photography and scanner imagery in the infrared portion of the spectrum have proven to be particularly useful in geologic applications because both solar reflectance and the inherent thermal radiation of materials may be observed. The terms used to designate the various infrared subregions are given in Table 4-2.

Photographic and scanner imagery from satellites in Earth polar orbit has been routinely available since the launching of Landsat 1 in 1972. The program provides coverage in four different wavelength regions or bands (Table 4-3) at a scale of 1:369,000.

The satellite track (Fig. 4-10) is the centerline of a 185 km-wide image swath that is displaced westerly on each orbit. The displacement is 160 km at the equator and 120 km at latitude 40° north and south, and the entire pattern is repeated each 18 days. Stereoscopic viewing is possible because of the partial overlap of images on successive days.

In 1984 the U.S. government announced plans to hand over its Landsat series of monitoring satellites to a private operator. At the same time, France launched the first of four planned commercial satellites termed SPOT (Systeme Pour l'Observation

Table 4-2.
Infrared (IR) Spectral Region

Wavelength in μm	Name	Remarks
0.7–300	IR	
0.7–0.9	Photographic IR	
0.7–3	Reflected IR	Primarily reflected solar radiation
3 –14	Thermal IR	Primarily inherent radiant temperature

Table 4-3.
Landsat Wavelengths

Band	Wavelength coverage (in micrometers)	Remarks
4	0.5–0.6	visible, green
5	0.6–0.7	visible, red
6	0.7–0.8	photographic, near 1R
7	0.8–1.	infrared

de la Terre) capable of a resolution of 20 meters in color and 10 in black and white, particularly useful for map scales of 1:50,000 to 1:100,000. The European Space Agency and Japan are preparing to follow suit, and the U.S.S.R. is offering images of 5-meter resolution for sale of all but its own and allies' territories.

Modern remote sensing systems all employ digital image processing for enhancement, comparison, quantification, and storage. Input data may be directly from the remote sensing scanner or scanned from a photograph or other image or object. basically, each picture element or *pixel* is assigned an x-y coordinate location on the image and a numerical value for its brightness. Pixel dimensions vary with image scale and program; they are about 30 × 30 m for Landsat and about 20 × 20 m for SPOT. Scales of brightness (*gray scales*) usually have 64 or 128 divisions.

Once stored, computer programs can call up wanted aspects of the digitized image. Counting of pixels in various levels in the gray scale provides percent area information, or gray scale values may be assigned colors. In this *false color* mode an area of known conditions brought to a distinctive color will provide a color reference to other areas of identical conditions (Fig. 4-11).

Imagery in the microwave region can be obtained by radar systems that illuminate

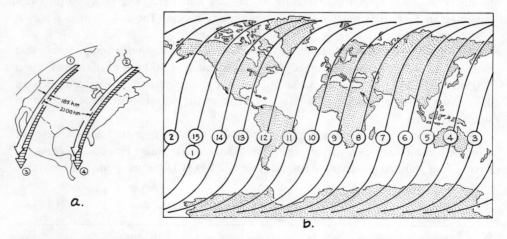

Figure 4.10. Landsat orbit paths. (a) Spacing and sidelap of tracks. Key: (1) Orbit N day, orbit N day M + 18. (2) Orbit N + 1 day M. (3) Orbit N day M + 1. (4) Orbit N + 1 day M + 1. (*Source:* Adapted with permission from *Remote Sensing: Principles and Interpretations* 2/e by F. F. Sabins, Jr. Copyright © 1978 W. H. Freeman and Company. Copyright © 1987 Remote Sensing Enterprises, Inc.) (b) Orbital paths for a single day. (*Source:* Based on *Landsat Users Handbook*, Goddard Space Flight Center, Doc. No. 76SDS-4528, NASA, 1976)

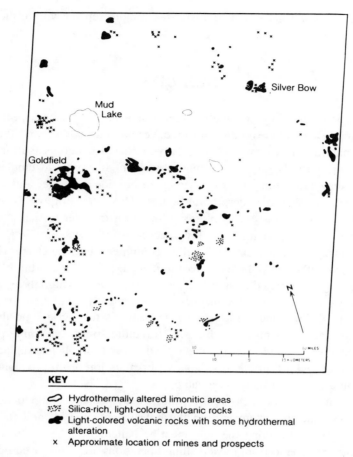

KEY

- ⬭ Hydrothermally altered limonitic areas
- ░ Silica-rich, light-colored volcanic rocks
- ⬛ Light-colored volcanic rocks with some hydrothermal alteration
- x Approximate location of mines and prospects

Figure 4.11. False color matching. (*Source:* From Rowan et al, *U.S. Geological Survey Professional Paper* 833, 1974)

the terrain and record radar echoes from an antenna mounted on an aircraft. The antenna axis in the usual configuration is side-looking, that is, angled downward from the side of the aircraft so it sweeps the terrain parallel to the flight path. The strength of the return signal is a function of surface roughness and absorption of radiant energy by various Earth materials. The product looks remarkably like a black-and-white photograph but differs significantly in that the longer radar wavelengths provide less resolution and shadows always fall away from the "camera." The system is particularly capable of enhancing linear features and can provide images through rain or cloud cover.

Aerial photography, side-looking radar, and satellite imagery are generally produced on rather small scales because they are products of public or quasi-public programs with many purposes to be served. Such imagery is excellent for geologic reconnaissance, but the scale is inadequate for detailed work. However, the availability of light aircraft, wide range of photographic emulsions, and hand-held scanning systems coupled with advanced television and microcomputer technology brings the availability of self-generated sophisticated remote sensing within the budgetary limits of even small exploration groups.

DRILLING

The drilling of holes by one or another means is a very common aspect of mineral resource work. Drilled holes are used to recover natural underground fluids (water, petroleum, natural gas, steam) or artificially fluidized materials (sulfur, dissolved metals), to inject fluids for underground storage or disposal, to emplace blasting charges in excavation, mining, or quarrying, to provide drainage, and to probe the nature, shape, and size of underground bodies. This latter purpose is often served in the final stages of a mineral exploration program in order to make the decision as to whether or not a potential orebody is economically viable.

There are many kinds and sizes of drilling equipment, but, except for flame-piercing methods, all either make holes by rotating an auger or abrasive bit or hammering a tool. Augering is restricted to such soft or unconsolidated materials as soil or coal, whereas both rotary and percussion drilling are typically employed for rock penetration. Rotary drilling and some percussion methods rely on a rigid drill stem to couple the driving mechanism and bit, which requires time-consuming pulling, uncoupling, recoupling, and lowering to recover the cored sample and replace the bit—hence the tall derrick of an oil rig. Cable-tool percussion drilling is done by lifting and dropping a heavy bit hung on a cable.

In most drilling a fluid such as air, water, or a specially compounded mud are circulated down the drill stem and up the hole to cool the bit and float out rock cuttings. In cable-tool drilling the hole must be bailed.

The samples recovered from most drilling operations are chips or cuttings brought to the surface (and possibly fractionated) by the returning drilling fluid where they may be caught on a screen for study. The more common exploration drilling, however, recovers a solid rock cylinder by employing an annular diamond bit backed by a core barrel, core being recovered after each few meters of advance. Diamond drilling has certain advantages also in the mobility of equipment and the ability to drill holes at any angle. The hole and core diameters of standard diamond drills are given in Figure 4-12.

Diamond drills should be operated by well-trained and technically competent personnel in order that drilling rates are maintained at high levels with minimum down time and complete core recovery. The location of the drill holes, however, is the responsibility of the geologist who must determine the position of collars, strike and dip of the hole, and its total length. The geologist must also examine and log the recovered core and do such mapping and interpretation as may be required.

Drilling to delineate subsurface conditions will be done by using some regular pattern of parallel or fanned holes spaced to provide the desired information. Drilling for exploration purposes, however, is not done on a preplanned grid, rather the information obtained from the completed holes is used to determine the next location.

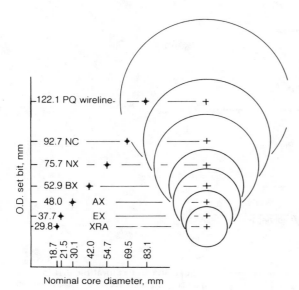

Figure 4.12. Standard diamond drill dimensions. (*Source:* From William H. Dennen and Bruce Moore, *Geology and Engineering.* Copyright © 1986 Wm. C. Brown Publishers, Dubuque, Iowa. All rights reserved. Reprinted by permission)

For example, consider the drilling sequence of a mineralized vein seen in outcrop at a single point; observation at the outcrop provides strike, dip, and width information, but drilling is needed to determine the lateral and vertical extent of the vein. The first hole should be located to test the vein at depth by drilling against the dip (Fig. 4-13a). The hole should be located to intersect the vein at an angle in excess of 45° and at a depth such that the square root of vein thickness times depth equals 100 or more. If depth extension is shown, the next holes should test the lateral extent of the vein by drilling at points along its strike to again provide a square root of thickness times length in excess of 100 (Fig. 4-13b). If all three holes make good intersections, the indicated volume is then at least 30,000 cubic units (Fig. 4-13c). The next sequence of holes should be planned to bracket the target by stepping out along strike and down dip at distances successively doubling the previous intersections. Finally, the body should be delineated by filling in the pattern at regular spacing.

GEOLOGIC EVALUATION

The general mechanisms whereby mineral accumulations may occur as normal consequences of geologic processes were reviewed in Chapter 2 and the means for their discovery are outlined above. Accumulations of minerals, however, do not necessarily constitute an ore because an ore deposit must be workable at a profit. The requirement of profitability brings many factors into play that will require careful assessment before a deposit will be worked. Purely geologic factors relating to the assessment of a deposit are discussed in the following pages and nongeologic factors in later chapters.

Of primary geologic concern in the economic assessment of a deposit is the total

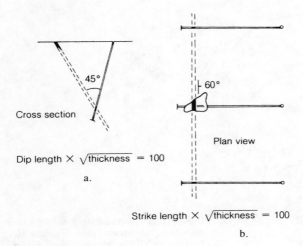

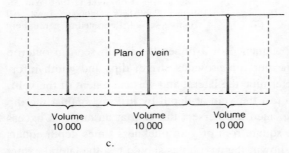

Figure 4.13. Exploration drilling sequence (see text) Key: (a) Cross section showing drill hole (solid lines) intersecting a vein (dashed) to test depth extension. (b) Plan view of drill holes intersecting a vein to test its strike extension. (c) Spacing of drill holes to block out ore. (*Source:* From William H. Dennen and Bruce Moore, *Geology and Engineering.* Copyright © 1986 Wm. C. Brown Publishers, Dubuque, Iowa. All rights reserved. Reprinted by permission)

amount of material present. This involves measurement of the volume of the deposit, determination of the tonnage (volume × bulk density), and establishment of the concentration of the substance of value per tonne by assay (grade or tenor). For bulk deposits of essentially "as is" material, such as limestone, glass sand, or gypsum, the assay may be made for impurities that reduce value.

The volume of a mineral deposit may be easy to calculate in the case of geometrically regular bodies such as tabular masses, cylinders, or spheres for which direct mensuration formulas are available:

The areas of some plane figures are:
  triangle = 1/2 of base × height
  rectangle = length × width
  circle = π × radius squared (π = 3.14159).
The volume of right prisms and cylinders are the area of their base × their height.
The volume of uniformly tapering solids such as cones and pyramids is 1/3 of the area of their base × their height.
The volume of a sphere is 4/3π radius cubed.

The volume of an irregular body is more difficult to obtain but may be found as the sum of those small regular units that comprise it. The easiest is to divide the volume into a series of parallel-sided plates of equal thickness but having irregularly shaped surfaces of different area. The areas of the surfaces may be found by any of several means

1. Measurement by planimeter, a mechanical device that integrates the area within a perimeter.
2. By the sum of areas of triangles.
3. By superimposing a regular grid and counting squares.
4. By weighing; an analytical balance may be used to weigh paper cutouts of areas.

and the average area of the bounding surfaces of a plate times its thickness is the plate volume; the sum of all plate volumes is the volume of the body (see Fig. 4-14).

The actual measurement of a body requires information in three dimensions. To obtain this, two-dimensional surface data are typically extrapolated to depth by drilling coupled with geologically reasonable assumptions. In this context it is particu-

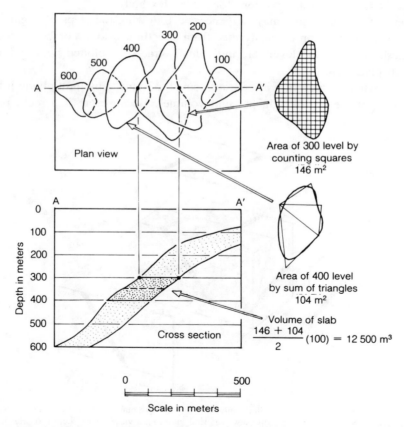

Figure 4.14. Measurement of irregular volumes. (*Source:* From William H. Dennen and Bruce Moore, *Geology and Engineering*. Copyright © 1986 Wm. C. Brown Publishers, Dubuque, Iowa. All rights reserved. Reprinted by permission)

larly useful to know a great deal about the nature of geologic boundaries and the shapes of rock masses.

Geologic boundaries range in character from knife-edge sharp to completely gradational over extended distances. The former are represented by readily mapped bedding surfaces, faults, intrusive contacts, or sharp-walled veins, and the latter by gradational sedimentary sequences, igneous aureoles, or wall-less veins. For such fuzzy contacts an arbitrary boundary or "assay wall" must be established by sampling or otherwise before a volume can be measured.

Calculation of volume is facilitated by the careful plotting of maps, construction of cross-sections, and the making of models. Unfortunately, three-dimensional information is not only difficult to obtain but often defies the simple representation needed to visualize and calculate. The usual means of representation is to generate a series of vertical or horizontal sections through the body that are parallel and evenly spaced. This series of plans or cross-sections may then be used as is, transferred to glass to make a sectional model, or transformed into an isometric block diagram. Figure 4-15 shows the terminology used for slices in different orientations and Figure 4-16 shows serial plans and sections for a complex body.

Isometric block diagrams may be generated following the steps of Figure 4-17. First the plan information is transformed from a rectilinear to a rhomboidal (60° and 120°) mesh and then the levels are individually traced or plotted with appropriate vertical displacement.

Volume, when measured, will be transformed to tonnage by multiplying it by the bulk density of the ore (weight per unit volume). Because ore is a rock, perhaps

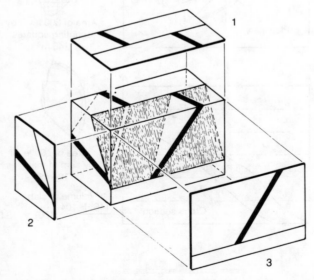

Figure 4.15. Block diagram of a faulted vein showing the orientation of standard views, variously called: (1) Top view, plan view or map view. (2) End view, side view, cross-section or elevation. (3) Front view or longitudinal section. (*Source:* From William H. Dennen and Bruce Moore, *Geology and Engineering*. Copyright © 1986 Wm. C. Brown Publishers, Dubuque, Iowa. All rights reserved. Reprinted by permission)

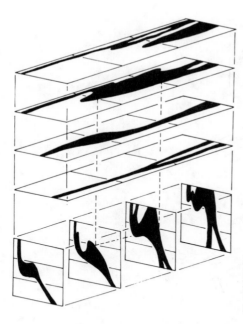

Figure 4.16. Serial plans and sections. (*Source:* From William H. Dennen and Bruce Moore, *Geology and Engineering*. Copyright © 1986 Wm. C. Brown Publishers, Dubuque, Iowa. All rights reserved. Reprinted by permission)

containing a significant percentage of voids, bulk density is properly determined on undisturbed samples or by weighing the material removed from a regularly shaped hole whose volume can be measured.

Determination of the ore value per unit weight or grade of both the various parts of a deposits and its overall average requires careful sampling and assay coupled with an appreciation of the variability of both natural content and analytic data. *Sampling* is a process whereby a small portion of a body is taken as representative of some larger volume and should be done in as regular and mechanical a manner as possible. Biased results because of contamination, salting (deliberate upgrading), or variable rock hardness must be guarded against.

Typical samples are rock chips, cuttings removed from shallow channels, excavated material from pits or trenches, or portions of diamond drill core. The pattern on which samples are collected may be random, but calculation is easier and the chance of preselection is reduced if a regular pattern is adopted. The spacing of samples must be such that any significant variation in grade will be detected, so it cannot be less than the width of the anomalous zone. The accepted dimensions may have to be found experimentally by comparing the assay results from nested samples with successively closer separation. Generally speaking, a larger number of smaller samples is to be preferred to fewer larger samples.

Large samples may be reduced in volume without change in quality by splitting or successive coning and quartering. The size of particles should also be reduced as samples become smaller. Crushing should be progressive and the rule is to have individual pieces so small that the removal or inclusion of the largest piece of best grade will not affect the final result.

Splitting is done with a device that divides the sample into two equal parts, one

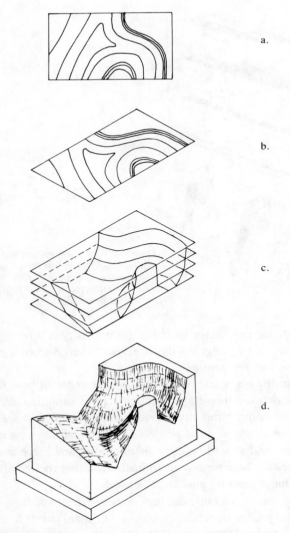

Figure 4.17. Steps in making an isometric block diagram. (a) Contour map or stack of horizontal sections. (b) Transformed from rectilinear to isometric (60–120°) coordinates. (c) Successive contour levels or sections displaced by a selected vertical interval, connected on visible edges, not drawn where hidden. (d) Worklines removed, connections made, block shaded. Scales, orientation, etc. added as desired. (*Source:* From William H. Dennen and Bruce Moore, *Geology and Engineering.* Copyright © 1986 Wm. C. Brown Publishers, Dubuque, Iowa. All rights reserved. Reprinted by permission)

of which is discarded. In coning and quartering a sample is poured in a conical pile onto a rubber sheet or canvas, the pile divided into quarters with a spatula, and alternate quarters discarded. The sample is then rolled by lifting alternate corners of the sheet to rehomogenize it, poured to form a cone, and quartered again. This sequence is continued until the wanted sample size is reached. The minimum weights of samples as a function of their grain size is given in Figure 4-18.

Samples, once obtained, are subjected to assay by an appropriate method. Choices

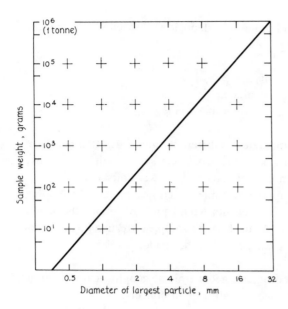

Figure 4.18. Permissible minimum weight of samples.

as to cost, speed, precision, and accuracy should be based on competent advice and may be done either by an in-house laboratory or by reputable assayers. In either instance, however, the explorationist should submit disguised samples of known content and resubmit portions of earlier samples at irregular intervals to guide the interpretation of the assay data. Note that no assay can be better than the sample it represents—and vice versa.

It is particularly important in the evaluation of analytic data to grasp the importance of its statistical variability. Accuracy or correctness, precision or repeatability, and bias or average difference are statistical terms used to describe analytic data. A simple way of visualizing them is to consider the bullet holes in the targets of Figure 4-19. High precision is shown by a small scatter and high accuracy by the (near) coincidence of the group average with the bullseye. Bias would describe the separation of the group average from the bullseye or between two groups by different shooters.

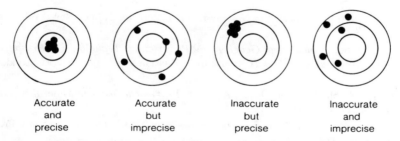

Figure 4.19. Precision and accuracy. (*Source:* From William H. Dennen and Bruce Moore, *Geology and Engineering.* Copyright © 1986 Wm. C. Brown Publishers, Dubuque, Iowa. All rights reserved. Reprinted by permission)

When the volume, bulk density, and grade of a deposit have been determined, its in-place value may be readily found:

Total tonnage = volume × bulk density.
Commodity tonnage = total tonnage × grade.
In-place value = commodity tonnage × value per tonne.

Because different commodities are marketed in different units, it is often necessary to convert the analytic data or calculated value from one set of units to another. (Appendix III provides some common conversions and market units.)

Many of the evaluative operations applied to mineralized ground are overlapping; for clarification the following sequence of events would probably mark the identification and evaluation of an orebody. At each stage samples are taken, maps and sections updated, and refinements of tonnage and grade made:

1. Location of mineralization by geological, geochemical, geophysical, or remote sensing means employed singly, sequentially, or in concert.
2. Large-scale geologic mapping and sampling of visually distinct materials.
3. Trenching (costeaning) or pitting to determine the geometry and obtain large samples.
4. Trial drilling of a few holes located to provide information as to the extent of the body, sampling of core, construction of cross sections.
5. Pattern drilling to clearly define the boundaries of the deposit, establish its internal variability, and provide core samples for assay.
6. Underground exploration shafts and tunnels constructed, mapped, and sampled.
7. Calculation of in-place value.

References

Beck, A. E. 1982. *Physical Principles of Exploration Methods*. London: Macmillan.
Brock, B. B. 1972. *A Global Approach to Geology*. Rotterdam: A. A. Balkema.
Cummings, J. D., and Wicklund, A. P. 1975. *Diamond Drill Handbook*, 3rd ed. Toronto: J. K. Smit and Sons.
Dobrin, M. B. 1976. *Introduction to Geophysical Prospecting*, 3rd ed. New York: McGraw-Hill.
Ginsberg, I. I. 1960. *Principles of Geochemical Prospecting; Techniques of Prospecting for Non-ferrous Ores and Rare Metals*. trans. from Russian by V. P. Sokoloff. London: Pergamon Press.
Hawkes, H. E., Webb, J. S., and Rose, A. W. 1979. *Geochemistry in Mineral Exploration*. New York: Academic Press.
Harris, D. P. 1984. *Mineral Resources Appraisal*. New York: Oxford University Press.
Lacy, W. C. 1982. *Mineral Exploration*. Stroudsburg, PA: Hutchison Ross.
Malyuga, D. P. 1964. *Biogeochemical Methods of Prospecting*. New York: Consultants Bureau.

Peters, W. C. 1987. *Exploration and Mining Geology*, 2nd ed. New York: John Wiley & Sons.

Reedman, J. H. 1979. *Techniques in Mineral Exploration*. Essex: Elsevier.

Sabins, F. F., Jr. 1986. *Remote Sensing*, 2nd ed. San Francisco: W. H. Freeman.

Siegel, F. R. 1974. *Applied Geochemistry*. New York: John Wiley & Sons.

Sharma, P. V. 1986. *Geophysical Methods in Geology*. New York: Elsevier.

Telford, W. M., Goldart, L. P., Sheriff, R. E., and Keys, D. A. 1976. *Applied Geophysics*. Cambridge: Cambridge University Press.

Ward, F. N., Lakin, H. W., Canney, F. C., and others. 1963. Analytical methods used in geochemical prospecting by the U.S. Geological Survey. *U.S. Geol. Survey Bull.* 1152.

Chapter 5
Classification of Mineral Deposits and Commodities

SCIENTIFIC CLASSIFICATION

At first glance the classification of ore deposits appears to be an unrewarding pedagogic process; pedagogic perhaps, but certainly not unrewarding. Good schemes of classification allow our knowledge to be organized in a condensed form that highlights similarities and differences in deposit genesis, age, location, or tectonic setting. A considerable transfer of information is possible when an undeveloped deposit can be placed in the same category as a well-studied one. Exploration is guided by an understanding of the processes of orebody genesis and the geologic/tectonic settings in which this occurs.

Discussions of mineral deposits depend ultimately on the basis of reference or scheme of classification in which they are viewed. Among the various parameters that may be employed are:

1. Distribution
    a. In time. Having features in common with other deposits of the same age; examples: Proterozoic banded iron formation, Permian evaporites.
    b. In space. Grouping by geographic location into mineral districts or belts containing related deposits; examples: Central African Copperbelt, Western Australia nickel province.
2. Genesis. Organization by process; examples: deposits produced by mechanical processes of concentration, deposits formed by chemical processes of concentration in magmas.
3. Tectonic setting. Correlation of deposit types with regions of distinctive tectonic style, examples: deposits of the ocean crust, surficial deposits of continental areas, deposits of continental seas, volcanic arc association.
4. Commodity
    a. Nature; examples: fuels, metals.
    b. Use; examples: gems, pigments.

The distribution of ore deposits in space and time is a useful collective means for grouping certain kinds of deposits or referring to a particular style of mineralization. The lack of certain knowledge about the timing of mineralization on one hand and uncertainty of district limits on the other, however, make such parameters of limited general applicability.

The review in Chapter 2 of geologic materials and processes provides the basis for a classification of mineral deposits according to the manner of their origin. Waldemar Lindgren in 1913 combined his observations with those of earlier workers to provide such a classification, which, with modifications, is widely used today. A modified version of Lindgren's scheme following Park and MacDiarmid (1975) is given as Table 5-1. The primary dependence of the organization of the table on geologic processes should be noted. A related scheme is given as Table 5-2.

Modern trends in mineral deposit study have placed major emphasis on the tectonic setting in which the various ore-forming processes are to be found. Although no generally accepted mineral deposit classification based purely on the tectonic regime has yet emerged, the principal settings and deposit types are well established and might be organized as shown in Table 5-3. Such a classification is particularly useful

Table 5-1.
Lindgren's Classification of Ore Deposits
(modified and abbreviated after Park and MacDiarmid, 1975)

I. Deposits produced by chemical processes of concentration. Temperatures and pressures vary between wide limits.
  A. In magmas, by processes of differentiation.
    1. Magmatic deposits proper, magmatic segregation deposits, injection deposits. Temperature 700° to 1500 °C; pressure very high.
    2. Pegmatites. Temperature very high to moderate; pressure very high.
  B. In bodies of rocks.
    1. Concentration affected by introduction of substances foreign to the rock (epigenetic).
      a. Origin dependent upon the eruption of igneous rocks.
      b. By hot ascending waters of uncertain origin, possibly magmatic, metamorphic, oceanic, connate, or meteoric.
      c. Origin by circulating meteoric waters at moderate or slight depth. Temperature up to 100 °C, pressure moderate.
    2. By concentration of substances contained in the geologic body itself.
      a. Concentration by dynamic and regional metamorphism. Temperature up to 400 °C; pressure high.
      b. Concentration by groundwater of deeper circulation. Temperature 0° to 100 °C; pressure moderate.
      c. Concentration by rock decay and residual weathering near surface. Temperature 0° to 100 °C; pressure moderate to atmospheric.
  C. In bodies of water.
    1. Volcanogenic. Underwater springs associated with volcanism. Temperatures high to moderate; pressure low to moderate.
    2. By interaction of solutions. Temperature 0° to 70 °C; pressure moderate.
    3. By evaporation of solvents.
II. Deposits formed by mechanical processes of concentration. Temperature and pressure moderate to low.

Table 5-2.
A Genetic Classification of Ore Deposits

Syngenetic Ore Deposits
(deposits formed as a part of the rock-forming process)

Magmagenic
  Igneous rocks
    aggregate, building, and decorative stone
  Disseminations
    kimberlite pipes (diamonds); alkaline igneous ring complexes (apatite, columbium, rare earths, uranium)
  Segregations
    Bushveld igneous complex, S. Africa (chromite, platinum);
    Insizwa, S. Africa (nickel)

Metamorphic
  Metamorphism of pre-existing ore
    Rammelsberg, Germany (copper, lead, zinc); Ducktown, TN (copper)
  Ores formed by metamorphism
    Spruce Pine District, NC (kyanite); St. Lawrence Co., NY (talc)

Sedimentational (stratiform deposits)
  Mechanical accumulations
    Detrital sedimentary rocks: Oriskany Sandstone, WVA (glass sand);
    coal underclay (fire clay)

Placer deposits
  gold: Mother Lode area, CA; Ballarat, Australia, Otago area, N.Z.
  tin: Kinte Valley, Malaysia; Rondonia, Brazil
  diamonds: West Africa; Angola; Venezuela; Brazil
  zircon: Australian beach sands
  platinum: Ural Mtns., USSR; Goodnews Bay, AL
  ilmenite: Nigeria; Blind River, Canada; Brazil; Kerala, India

Evaporites
  salt, potash: Zechstein Formation, Europe; Prairie Formation, western Canada
  borates: Searles Lake, CA; Argentina
  gypsum: Michigan, Iowa, Texas, California; Nova Scotia

Chemical precipitates*
  ironstone: Silurian Clinton ore, U.S.
  banded iron formation (BIF): see map, Figure 2–14
  manganese: sea floor nodules, western Pacific; Nsuta, Ghana
  phosphatic sediments: Phosphoria Formation, northwest U.S.
  limestone: widespread

Volcanic-exhalative*
  Cyprus-type: Cyprus; Oman (manganese)
  Kuroko-type: Hokuroka District, Japan; Rio Tinto, Spain, Kidd Creek, Ontario (pyrite and base metal sulfides)
  Besshi-type: Shikoko Island, Japan; Taiwan (base metal sulfides)
  Sublimates

Bacterial*
  sulfur: in salt dome cap rock, Gulf Coast
  base metal sulfides: Mt. Isa, Australia; Hockley Dome, LA

Weathering
  manganese: Amapa, Brazil; Madhya Pradesh, India; Postmasburg, S. Africa
  barite: MO, AK

Table 5-2 *continued.*
Syngenetic Ore Deposits
(deposits formed as a part of the rock-forming process)

bauxite: AK; Weipa, Australia; Guinea; Jamaica
iron: Lake Superior District; Carajas, Brazil; Mount Newman, Western Australia; Nimba, Liberia

Epigenetic Ore Deposits
(ore introduced into a pre-existing host rock; often requires ground preparation)

Magmagenic
  Residual liquid injection
    Adirondacks, NY; Kiruna, Sweden; Pea Ridge, MO (iron)
  Immiscible liquid injection
    Vlackfontein, S. Africa (sulfides)
  Pegmatites
    eastern U.S., see map in Figure 2-7; Bikita, Zimbabwe; Madagascar, Malagasy Republic; Bihar and Madras, India; Minas Gerais, Brazil; (lithium, cesium, beryllium, mica)
Pyrometasomatic
  copper, tin, tungsten: Cornwall, England; Pasto Bueno, Peru; Panasqueira, Portugal; King Island, Australia; Mt. Bischoff, Australia
  iron: Larap, Philippines; Bukit Ibam mine, Malaysia; Cornwall, PA
Diagenetic (products of interactions between primary sediments and pore waters derived essentially from their dewatering)
  red bed-copper association: Kupferschiefer, Germany; Central African Copperbelt; White Pine District, MI
  sandstone uranium-vanadium-copper association: Colorado Plateau; Great Slave Lake, Canada; Beaufort Basin, S. Africa
Hydrothermal
  Sediment-hosted lead-zinc deposits
    Carbonate-hosted Mississippi Valley-type: East-central U.S.; Pennines, England; Pine Point, Northwest Territories
    Carbonate-hosted Irish-type
    Shale-hosted type: Sullivan, B,C.; Mt. Isa, Australia; McArthur River, Australia
  Veins, stockworks, saddle-reefs, breccia fillings, etc.: Pachuca, Mexico (silver); Bendigo, Australia (gold); Mascot, TN (zinc)
  Replacements
    Bisbee, AZ (copper); Kirkland Lake, Canada (gold)
  Alteration
    Supergene enrichment: Ely, Nevada; Chuquimata, Chile (copper); Mount Morgan, Queensland (gold)
  Bacterial action

*indicates a significant bacterial involvement possible

because, if the tectonic setting is known, an immediate tabulation of the kinds of exploration targets to be anticipated is available. The main complication is that the setting of concern is not always that of today but of the time of mineralization.

  Classification of the thousands of known ore deposits is a task far beyond the scope of this book. It is, however, strongly recommended that ore deposits visited in the field, discussed in the classroom, or encountered by reading be located in one or more of the schemes described or into a personally devised organization.

Table 5-3.
Tectonic Regimes of Ore Deposits

Ocean basins	
Passive regions	Sea floor ferromanganese nodules
Hot spots	
Spreading centers	Podiform (alpine-type) chromite in ophiolite suite association
	Stratiform volcanic-exhalative massive sulfides and bedded ferromanganese deposits
Cratons	
Intrusive bodies	Alkaline ring complexes, carbonatites, and peridotite pipes. Pegmatites
Astroblemes	Sudbury basin, Ontario (?), Bushveld Igneous Complex, South Africa (?)
Intracratonic basins	
Syngenetic deposits	Banded iron formation, ironstone deposits (Clinton ores, minette), bog iron ores, bedded manganese deposits, placers
Epigenetic deposits	Carbonate and shale-hosted lead-zinc deposits, uranium, vandium, copper deposits in sandstone, red bed-copper association
Epicontinental seas	Limestone, evaporites, phosphorites, uraniferous black shale
Residual orebodies	Bauxite, nickel, manganese, iron, phosphate, and barite deposits
Rift zones	Stratabound tungsten-tin deposits, stratiform sulfide deposits, tin-bearing granites, pegmatites
Convergent zones	
Continent-continent	Tin-bearing anatectic granites, uraniferous granites
Ocean-Continent	
Magmatic arcs	Copper-molybdenum porphyries, Kuroko-type stratiform sulfide deposits
Back arcs	Epithermal vein deposits of iron, tin, antimony, precious, and base metal ores
Outer arcs	Tin-copper-arsenic deposits

MINERAL COMMODITY CLASSIFICATION

Scientific schemes for the classification of mineral deposits are particularly useful in the understanding and discovery of mineral resources, but the markets for these resources are conditioned by the uses to which the material is put and not its origin. Table 5-4 is a listing of minerals and rocks arranged by their use, which, although far from complete, suggests the wide range and importance of geological materials in a technological society.

The marketplace has no interest in the scientific classification of mineral deposits or generalizations regarding the use of raw materials; rather, it deals in specific commodities. The terminology that has grown up for this purpose is a hodge-podge of

Table 5-4.
Usage Classification for Mineral Commodities

Fuels
    Fossil: petroleum, natural gas, oil shale, tar sand, coal, lignite
    Nuclear: ores of uranium, thorium
Nonmetallics
    Petrochemicals: petroleum, natural gas, asphalt, coal
    Construction materials (fill, aggregate, dressed stone): most rocks
    Mortars and cements: limestone, dolostone, gypsum
    Land plasters and fertilizers: limestone, dolostone, phosphate, gypsum
    Glass and ceramics: glass sand, clay, feldspars, nepheline
    Abrasives: quartz, corundum, garnet, diamond
    Pigments: ores of titanium, iron, cobalt, cadmium, copper
    Refractories: ores of magnesium, asbestos, alumina, chromite
    Fluxes: quartz, fluorite, borates, limestone
    Nuclear materials: beryllium, graphite
    Lubricants: graphite, molybdenite
    Absorbents: fullers earth, clay
    Mordants: chromite, hematite
    Basic chemicals: halite, sulfur
    Gemstones: diamond, corundum, garnet, quartz, beryl
    Pyrotechnics: strontium
    Miscellaneous: barite (drilling mud), rare earths (sparkers), molding sands, soil stabilization
Metallics
    Ferrous metals: ores of iron, cobalt, chromium, vanadium, tungsten, manganese
    Base metals: ores of copper, lead, zinc, tin, molybdenum
    Light metals: ores of aluminum, magnesium, titanium, beryllium, lithium
    Precious metals, ores of gold, silver, platinum group elements
    Miscellaneous minor metals: ores of gallium, indium, cadmium, germanium, mercury, antimony, bismuth, rare earth elements

rock, mineral, chemical element, chemical compound, and usage terms having no rationality except in their general acceptance. Tables 5-5 and 5-6 present the categories used by the U.S. Bureau of Mines and by the periodical Industrial Minerals.

Table 5-5.
Mineral Product Categories Employed by the U.S. Bureau of Mines

Chemical element or compound	Mineral	Rock	Usage
aluminum	asbestos	bauxite	abrasive materials
antimony	barite	coal	cement
beryllium	clays	diatomite	ferroalloys
bismuth	feldspar	nepheline	gemstones
boron	fluorspar	syenite	pigments
bromine	graphite	aplite	
cadmium	gypsum	iron ore	
calcium and calcium compounds	kyanite	peat	
	mica	perlite	
carbon black	rare earth minerals	phosphate rock	
chromium	pyrites	pumice and volcanic cinder	
cobalt	talc		
columbium and tantalum	vermiculite	pyrophyllite	
copper		salt	
gallium		sand and gravel	
gold		stone	
helium		soapstone	
iron and steel			
iron oxide			
lead			
lime			
magnesium		**Miscellaneous**	
magnesium compounds			
manganese		natural gas	
mercury		natural gas liquids	
molybdenum		petroleum and petroleum products	
nickel			
nitrogen			
platinum group metals		slag-iron and steel	
potash		minor metals	
rhenium		minor nonmetals	
silicon			
silver			
sodium and sodium compounds			
sulfur			
thorium			
tin			
titanium			
tungsten			
uranium			
vanadium			
zinc			
zirconium and hafnium			

Table 5-6.
Mineral Product Categories Employed by "Industrial Minerals"

Chemical element or compound	Minerals	Rocks	Usage
alumina	asbestos	bauxite	abrasives
antimony	attapulgite	aplite	iron oxide pigments
borates	barite	ball clay	
bromine	boron minerals	bentonite	
calcium carbonate	chromite	flint clay	
iodine	feldspar	nepheline syenite	
manganese	fluorite	perlite	
nitrate	graphite	salt	
phosphates	gypsum	silica sand	
potash	kaolin	slate	
soda ash	leucoxene		
sulfur	lithium minerals		
	magnesite		
	mica		
	olivine		
	pyrophyllite		
	rare earth minerals		
	rutile		
	sillimanite minerals		
	strontium minerals		
	talc		
	vermiculite		
	wollastonite		
	zircon		

References

Hutchinson, C. S. 1983. *Economic Deposits and Their Tectonic Setting.* New York: John Wiley & Sons.

Industrial minerals. *Metall. Bulletin Ltd.,* London.

Jensen, M. L., and Bateman, A. M. 1979. *Economic Mineral Deposits.* New York: John Wiley & Sons.

Lindgren, W., 1913. *Mineral Deposits.* New York: McGraw-Hill.

Mitchell, A. H. G., and Garson, M. S. 1981. *Mineral Deposits and Global Tectonic Settings.* New York: Academic Press.

Minerals yearbook. U.S. Bureau of Mines, Washington.

Park, C. F. Jr., and MacDiarmid, R. A. 1975. *Ore Deposits,* 3d ed. San Francisco: W. H. Freeman.

Tarling, D. H., ed. 1981. *Economic Geology and Geotectonics.* New York: John Wiley & Sons.

U.S. Geological Survey, 1986. *Mineral Deposit Models.* U.S. Geological Survey Bulletin 1693. Washington.

Part II

Extraction and Milling

Chapter 6
Mining

INTRODUCTION

Mining is the general term used for the separation of an ore from the ground, and *milling* refers to the processing of the run-of-mine ore into one or more marketable products. The combination of methods selected for the recovery of a given ore should, of course, be the optimum one to generate maximum profit. This is a particularly difficult task because it must include choices of mine geometry, machinery, operational plans, milling methods and equipment, and marketing strategy within a predicted economic framework; it will almost certainly involve a number of trial combinations in the planning stages.

The separation of ore from the ground by mining, quarrying, dredging, or through bore holes followed by either removal of the ore from gangue (waste rock) or impurities from bulk comodities are the special province of the mining engineer and mineral dresser, respectively. They cannot work independently from one another and must interact both with geologists and executives making business decisions throughout the life of a mine.

Knowledge of the quantity, geometry, physical properties, grade, and potential market of the ore in a deposit is fundamental to decisions relating to the optimum manner and rate of ore production to be employed. From these and other economic considerations discussed later, a planned annual production rate is established. This may be simply the total recoverable tonnage divided by a predetermined lifetime (time to exhaustion) of the deposit, or will more likely be dictated by the choice of mining method and mill capacity vis-à-vis a desired level of return on investment.

It should always be kept in mind that the determination that a deposit apparently contains ore of sufficient quality and quantity to warrant its exploitation does not automatically lead to a decision to bring it into production. Extraction and beneficiation of the ore must be considered in terms of future cost and markets, and the balance sheet may not always be favorable.

EXTRACTION METHODS

General Principles

The planning and conduct of a mining program entails the three-dimensional visualization of a continuously changing workplace that must be safe, well drained and ventilated, have appropriate power and transportation systems, and be able to generate ore at a predetermined rate and quality. Planning in such a way as to provide these needs and obviate potential and costly hazards is at least as important as the actual production activity. A well-planned and functioning mine is an accomplishment equivalent or surpassing more visible products of engineering science.

Because of their different modes of formation, orebodies tend to be two-dimensional sheetlike or tabular bodies, often more or less crumpled; linear blades or rods; or irregular three-dimensional masses. Mining of such bodies will require different approaches according to their thickness, attitude, amount of cover, and strength of the ore and its enclosing rocks.

Mine openings must be large enough to provide access for miners and machines. If the "mining width" is greater than the orebody thickness, some wall rock must be excavated with the ore, thus reducing its grade. Alternatively, if ore thickness is greater than can be safely removed in a single operation, several lifts must be made. Flat-lying orebodies may be mined on a single level, whereas gravity can be used in steeply dipping bodies to bring ore down dip to transportation levels paralleling the strike. The strength of enclosing wall rock determines the size of openings, nature and amount of support that must be provided, and may fix the mining plan. Weak rocks, for example, may be deliberately caved.

Basic to mine planning is the nature of the ore and the shape, size, attitude, and cover of the orebody because these dictate whether the ore may be extracted through bore holes, from open pits, or by underground methods. Bore hole recovery is limited to underground fluids or substances that can be made so by heat or solvents and has the advantages of low cost and minimal surface disturbance that are offset by low percentage recovery. For solid ores, open pit methods are often preferred because of their relatively lower operating costs, greater percentage recovery, and safer working conditions when compared with underground mining. Open pit methods do, however, cause much greater disturbance of the surface and require extensive postmining reclamation.

The range and variability of ore extraction methods needed to handle the plethora of mining situations may be appreciated by an examination of Table 6-1, which categorizes the more important means of recovery. In the selection of the method of mining for a given orebody, usually several methods are adapted to accommodate the conditions at the site. The resultant plan is seldom ideal from all points of view, but should represent the best compromise between conflicting factors.

Rock Blasting

A detonation of sufficient strength within a volume of rock will blow out a conical mass of broken material whose axis is from the point of detonation to the nearest

Table 6-1.
Mining Methods

Surface Mining
    Open pit (open cast, strip)
    Quarrying
    Hydraulic mining
    Dredging
Combined surface and underground
    Glory hole
Underground stoping
    Naturally supported stopes
        Open stopes
        Open stopes with pillars
    Artificially supported stopes
        Shrinkage stopes
        Cut and fill stopes
        Stulled stopes
        Square set stopes
    Caved stopes
        Block caving
        Sublevel caving
        Top slicing
        Combined methods
Bore hole recovery
    Auguring
    Coaxial holes
    Well field

---

free face. The apical angle of the cone is 90° for homogeneous material, increasing as the rock strength in the direction of the axis becomes less than that across it. The charge needed to break and displace the rock in the cone is a function of the explosive used, rock strength, and the distance to the free face; more explosive breaks the rock more finely and shatters a surrounding volume, but does not displace any more.

Only in rare circumstances is the explosive charge concentrated at a point. It is usually positioned by the use of drilled holes and is a cylinder of finite length and fixed attitude. Breakage from its detonation is thus in the form of a blunt wedge of rock. Obviously, the selection of explosive type and amount (a function of hole diameter), and the spacing and attitude of holes for charging in relation to the amount and particle size of broken rock to be produced calls for special skills on the part of the blaster. The best type, amount, and placement of explosive in a given case can be determined only by experience. Because the subsequent treatment of the material must be considered, charges are calculated to provide particle sizes no greater than can be handled by mucking machines or primary crushers without producing excessive amounts of fines. Holes for blasting are usually made by air-operated drilling equipment because of its relatively low cost and high efficiency. An added advantage underground is that the exhaust helps ventilate close working areas.

The volume of rock displaced by a given charge is approximately equal to the number of free faces. In tunneling and shaft sinking, only one free face is present and several independent blasts are required to advance. Charges in holes nearer the center and inclined inward toward the axis of the bore are detonated first and the conical space produced is then trimmed to size by charges in peripheral holes. Usually this is done by timed detonations a few milliseconds apart.

The blasting down of benches in open pit mines and most underground stoping layouts takes advantage of the presence of two free faces to improve excavation efficiency. In typical bench blasting, holes to the depth of the bench height are drilled three-quarters to one bench height back from the lip and the same distance apart. Charges in these holes are usually fired by timed detonation creating a ripple effect in order to obviate damage to structures.

Open Pit Mining and Quarrying

Given the choice, an orebody will, with rare exceptions, be worked open to the sky. Under these circumstances standard civil engineering earth and rock moving techniques and equipment may be adapted to extraction of the ore and higher productivity and recovery together with lower costs and greater safety can be achieved as compared with underground mining. Quarrying is the term used for an open pit mining operation whose product is usually some rock material exploited for itself alone such as granite, marble, slate or the like. The principal distinction between some quarrying and open pit mining is the deliberate removal in quarrying of blocks or dimension stone that requires the use of special excavation techniques.

More or less overburden must usually be stripped away in order to expose an orebody for open pit mining. The removal and replacement of this overlying soil or rock may represent a considerable fraction of the mining cost and the economic viability of a deposit is controlled by how much barren material must be included in calculations of grade.

For flat-lying, bedded deposits such as coal or limestone, the thickness of overburden, which is to be removed per unit thickness of ore to be mined, is called the overburden ratio. For example, the economic limit for stripping flat-lying coal seams in Kentucky is currently 12:1, and so a 1-meter-thick seam covered by 10 meters of overburden may be won by open cast mining, whereas the same seam under 15 meters of cover is uneconomic.

The walls of an open pit mine must be maintained at a safe angle, which for most operations is close to 45°, but for unconsolidated material might be 30° or less. At an average wall angle of 45°, each unit in depth requires the excavation of two units in pit breadth (Fig. 6-1). Obviously, this is a losing proposition for thick overburden or deep orebodies and in many instances open pit mines have been transformed into underground operations extending downward from the pit floor in order to exploit the deeper ore. The development of shafts and tunnels beneath an open pit to facilitate the removal of the orebody shifts the focus of activities below ground and transforms the operation into a glory hole mine.

The limiting depth for open pit mining may be found from the geometry of the orebody and grade of the ore. For example, in Figure 6-2 an orebody might be won

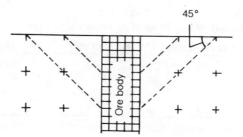

Figure 6.1. Breadth-depth relations in an open pit. (*Source:* From William H. Dennen and Bruce Moore, *Geology and Engineering*. Copyright © 1986 Wm. C. Brown Publishers, Dubuque, Iowa. All rights reserved. Reprinted by permission)

from a successively deepened pit. Assuming the ore grade to be constant, it can be seen that the average mined grade must continuously decrease because of dilution by wall rock as shown in Table 6-2. The quality of mineral matter mined may be held constant by selective mining, but the average grade of the deposit on which economic judgments are based decreases. The ratio of the tonnage of barren overburden and wall rock removed to the ore mined is the stripping ratio.

To clarify, if ore, gangue, and overburden plus wall rock are respectively A, B, and C tonnes, then

$A/(A + B) \times 100$ = ore grade.
$C/(A + B)$ = stripping ratio.
$A/(A + B + C)$ = mined grade.

Pits are usually developed as a series of steps or benches up to 15 meters high, which spiral around the pit to serve as working and transportation platforms. Several bench faces of differing grade will probably be worked simultaneously to allow blending of the run-of-mine ore to a mill feed of constant quality. A generalized plan of an

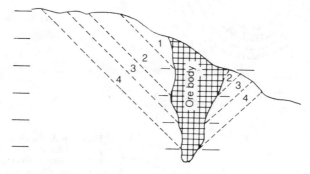

Figure 6.2. Volumes of ore and waste in a deepening pit. (*Source:* From William H. Dennen and Bruce Moore, *Geology and Engineering*. Copyright © 1986 Wm. C. Brown Publishers, Dubuque, Iowa. All rights reserved. Reprinted by permission)

Table 6-2.
Ore and Waste Production from Pit in Figure 6-2.

Slice	Ore Volume	Waste Volume	Waste: Ore Ratio[1]
1	25	21	0.84
2	40	73	1.83
3	30	80	2.67
4	20	127	6.35

[1]Equals the stripping ratio if the density of waste and ore is the same.

open pit mine is shown in Figure 6-3. In operation, successively lower benches are widened by blasting and ore removal so the pit is continuously widened and deepened. Eventually depths may be reached such that underground operations become more economical. Transition to deep mining is usually accomplished by the development of either an ore pass to a collection pocket with skip or belt transfer to the surface (Fig. 6-4a), or development of a glory hole by caving from below (Fig. 6-4b). The reverse may also occur. Underground mining operations for molybdenum at Climax, Colorado, were transformed to a glory hole mine and then to an open pit; the underground caving operation in copper ore at Ruth, Nevada, was completely excavated in the expanding Ruth pit.

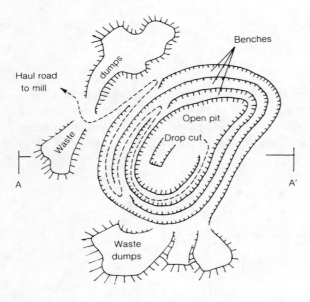

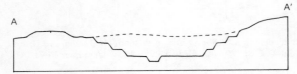

Figure 6.3. Open pit mine section and plan. (*Source:* From William H. Dennen and Bruce Moore, *Geology and Engineering.* Copyright © 1986 Wm. C. Brown Publishers, Dubuque, Iowa. All rights reserved. Reprinted by permission)

Mining

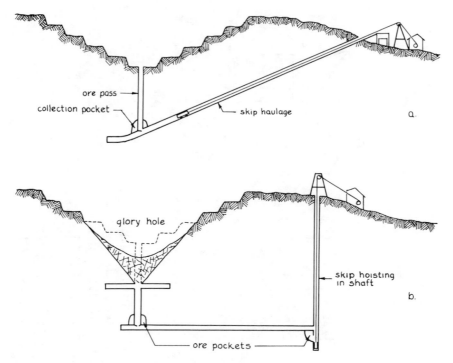

Figure 6.4. Transition to deep mining from an open pit.

Dredging

Unconsolidated ores under shallow water or in areas having a high water table can be recovered by dredging. Typically, dredged ores are those of high unit value such as gold or tin ore found in extensive placer deposits.

A dredge is essentially an integrated floating mining and processing plant that makes its own moving pond as it works. Ore is brought up by either suction or a bucket line, processed aboard, and the waste dumped behind. The digging device and waste carrier are mounted fore and aft and adjustable cables to the bank allow positioning of the dredge. The general arrangement is shown in Figure 6-5.

Underground Mining

The layout of an underground mine is necessarily more complex than that of an open pit, and being designed for a particular orebody, it will vary from mine to mine. For all of the great diversity of layout, however, some principles are common to all underground operations and mining techniques have been basically similar for many years; undisturbed ore or rock is left in place as pillars or ribs to safeguard the integrity of mine openings; gravity is used whenever possible to reduce handling

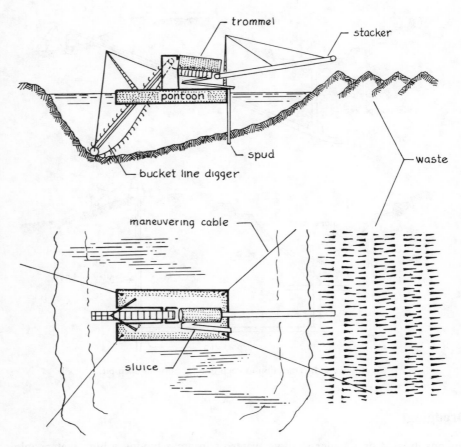

Figure 6.5. Floating dredge.

costs (ore removed from below by stoping or caving, tunnels slope downward from working faces to facilitate ore transport and drainage (ventilation is from below upward); headroom is usually kept small to reduce the danger of rock falls from the back or roof; and artificial support by timbering, roof bolting, packwalling, or backfilling is provided whenever there is danger of roof collapse or falling rock.

The guiding factors in the choice of an underground mining method are the shape and attitude of the orebody and the mechanical strength of the ore and the wall rocks (Table 6-3). The initial stages of mine development, shared by all methods, are the provision of an access system to the orebody analogous to the scaffolding used by builders. A tunnel, inclined or vertical shaft from the surface, horizontal crosscuts to the orebody, and drifts along it as shown in Figure 6-6 are the primary access openings. They are usually placed beneath the orebody in the footwall where they will be unaffected by breakage and settlement of overlying rocks as mining proceeds. A glossary of terms for mine openings is provided in Table 6-4.

Extraction of ore from within the body takes place in rooms or stopes connected more or less directly to the drifts (Fig. 6-7). This general plan is very flexible and

Mining

Table 6-3.
Selection of Underground Mining Methods

Mining method	Orebody Shape			Dip		Ore Strength		Wallrock Strength	
	Thin tabular	Thick tabular	Massive	Flat	Steep	Weak	Strong	Weak	Strong
Room and pillar	●	●		●			●		●
Casual pillar	○	○		○	○	○	○		○○
Longwall	●			●		●	●	●	
Open stoping	○	○○			○○○	○	○○	○○	○○
Stulled stoping	●				●	●	●	●	●
Square set stoping		○○	○	○	○	○○○	○	○○	○○○
Shrinkage stoping		●	●		●	●●			●●
Cut and fill stoping	○○		○	○○		○○○		○	○○○
Sublevel stoping		●●	●		●●	●●●		●	●
Top slicing	○	○○	○	○	○	○○○	○○	○	○○
Sublevel caving		●●	●	●	●	●●●	●●	●●●	●
Block caving		○	○	○		○	○		○○

can accommodate bodies with different width and dip, becoming a room and pillar arrangement when horizontal (see Fig. 6-14). The interconnection of stope and haulage drift is not always simple as, for example, in the Mount Isa copper mine, Australia (Figure 6-8).

As mining proceeds to greater depths or the strengths of rock and ore decrease, the size of openings that can be maintained also decreases. Internal support in stopes

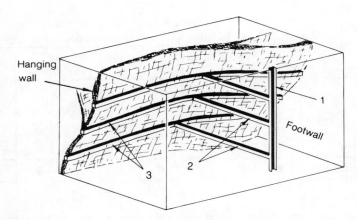

Figure 6.6. Access to the orebody. Key: (1) Vertical Shaft (May be an inclined or horizontal adit depending on topography) (2) Levels or crosscuts. (3) Drifts (Parallel to the orebody and either below or within it) (*Source:* From William H. Dennen and Bruce Moore, *Geology and Engineering.* Copyright © 1986 Wm. C. Brown Publishers, Dubuque, Iowa. All rights reserved. Reprinted by permission)

and other openings may be provided by artificial means, undisturbed ore may be left as pillars (perhaps to be recovered later), or stopes may be kept essentially full of broken ore or waste.

Broken ore occupies 30 to 50% more volume than the same rock unbroken and shrinkage stoping takes advantage of this fact. Stopes are maintained nearly full of broken ore with the necessary working space at the top of the stope provided by careful drawdown. Stope access may be by sublevels from a raise in a pillar, cribbed (timbered) raise from the level below, or a winze from the level above (Fig. 6-9).

Stopes may also be maintained in a nearly full condition by using waste rock from mining or milling. This is cut and fill stoping (Fig. 6-10) and in modern mines tends to be the automatic first choice for a feasibility study since it is the most versatile of underground mining methods; development costs are relatively low, massive as well as tabular bodies may be stoped, pillars and ore stringers into the country rock can be recovered, the design adapts well to mechanization, and underground waste disposal obviates environmental problems.

For large, low-grade orebodies or mining in mechanically weak ground, it may be advantageous to initiate and maintain a deliberate caving of the ore. This is done by removing support from one layer at a time—sublevel caving—or from a large block of ore in block caving (as shown in Figs. 6-11, 6-12). Advantages are low mining costs coupled with high production rates offset by high initial development outlay, dilution of the ore by wall rock and capping, and disturbance of the surface.

Mining may be done either in advance from or retreat to the point of access to the mine. Most commonly it is done by advance from the shaft or adit to a perimeter marked by the limit of the orebody or property lines followed by a retreat during which supporting pillars are removed or robbed. This procedure allows an ore flow and hence cash flow to start early. Mining in retreat, whether from the outset of mining or following advance to the perimeter, removes internal support with the

Table 6-4.
Glossary of Terms for Mine Openings

Adit	A horizontal gallery driven from the surface and giving access to an orebody that is worked through it; used sometimes solely for drainage or ventilation or both. The term *tunnel* is frequently used in place of adit and has the same significance.
Crosscut	A horizontal gallery driven at right angles to the strike of a vein. When driven at an angle to the vein and across it the same term is applied.
Drift	A horizontal gallery driven along the course of a vein. The terms *footwall drift* and *hangingwall drift* are used when galleries are driven in the footwall and hangingwall respectively.
Incline	An excavation of the same nature as a shaft and used for the same purposes but driven at an angle from the vertical.
Level	All of the horizontal workings tributary to a given shaft station are collectively called a *level*. Levels are designated 100 ft., 200 ft., etc., the vertical depth from the surface determining.
Main level	Where the ore mined from several levels is hauled upon one level to the shaft, this level is designated as the "main level." The term *main haulage* is used in the same sense.
Manway	Passages either vertical or inclined for the accommodation of ladders, pipes, etc.
Ore pass or chute	Vertical or inclined passageways for the downward movement of ore.
Ore pocket	Opening below a level for the temporary storage of ore at, and connected to, the shaft through which the ore is hoisted.
Raise	An excavation of restricted cross section, driven upward from a drift and in the orebody. It is used as a manway, timber chute, waste chute, ore chute, or for ventilation.
Shaft	A vertical excavation of restricted cross section and of relatively great depth used for access and working. Often compartmented.
Station	Junction of a level with a shaft.
Stope	An excavation underground, other than development workings, made for the purpose of removing ore.
Sublevel or intermediate level	A level driven from a raise or manway and not connected directly with the working shaft.
Underground shaft or incline	A shaft or incline that is driven from underground workings and not in the vein or orebody.
Winze	An excavation of restricted cross section driven downward from a drift and in the orebody. It is used for the same purposes as a raise.

From G. J. Young. *Elements of Mining*. © 1932 McGraw-Hill Book Co., N.Y. Reprinted by permission of the McGraw-Hill Book Co.

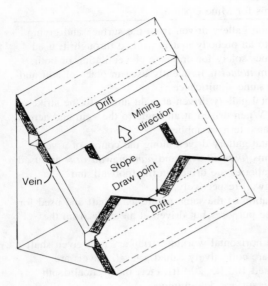

Figure 6.7. Arrangement of drift and stope. (*Source:* From William H. Dennen and Bruce Moore, *Geology and Engineering*. Copyright © 1986 Wm. C. Brown Publishers, Dubuque, Iowa. All rights reserved. Reprinted by permission)

consequent likelihood of surface subsidence. If this is of importance, the costs of backfilling must be weighed against the value of ore to be recovered by robbing of pillars.

Plans for mining in advance and in retreat are shown in Figures 6-13 and 6-14. These are for coal mines in this instance, but the general methods are applicable to any tabular, flat-lying orebody.

Miscellaneous Extraction Methods

Borehole extraction. A few commodities exist in the ground in gaseous, liquid, or liquifiable form and may be extracted through boreholes. Included here are, of course, petroleum and natural gas as well as helium, carbon dioxide, and steam.

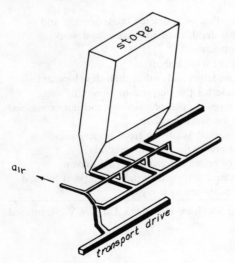

Figure 6.8. Drift-stope interconnection in a sublevel stoping operation, Mount Isa, Australia.

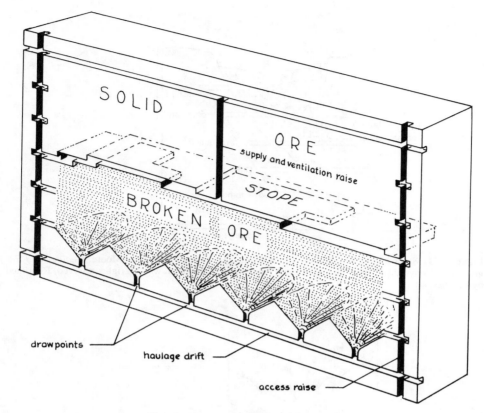

Figure 6.9. Layout of a shrinkage stope.

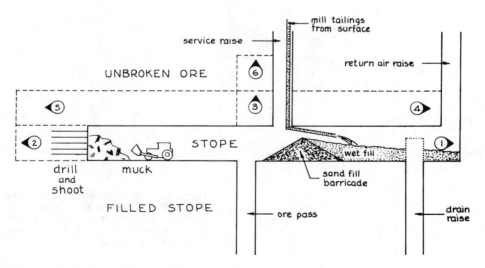

Figure 6.10. Cut and fill stoping; the sequence of operations is indicated by the arrowed numbers.

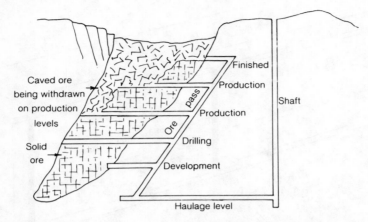

Figure 6.11. Sublevel caving. (*Source:* From William H. Dennen and Bruce Moore, *Geology and Engineering*. Copyright © 1986 Wm. C. Brown Publishers, Dubuque, Iowa. All rights reserved. Reprinted by permission)

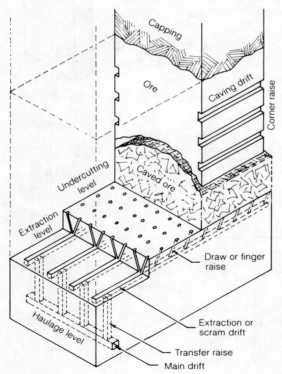

Figure 6.12. Block caving. (*Source:* From William H. Dennen and Bruce Moore, *Geology and Engineering*. Copyright © 1986 Wm. C. Brown Publishers, Dubuque, Iowa. All rights reserved. Reprinted by permission)

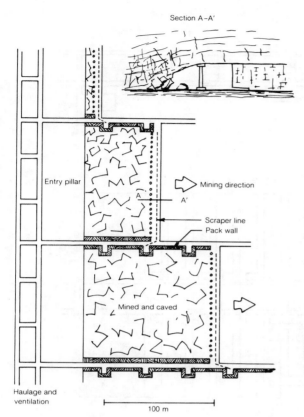

Figure 6.13. Mine plan, longwall advancing. (*Source:* From William H. Dennen and Bruce Moore, *Geology and Engineering.* Copyright © 1986 Wm. C. Brown Publishers, Dubuque, Iowa. All rights reserved. Reprinted by permission)

Additionally, easily melted native sulfur, soluble rock salt, and several metals, notably copper and uranium, which may be dissolved from their ore minerals by appropriate reagents, may be recovered through boreholes. Recovery may be through single coaxial holes or by the installation of a field of injection and recovery wells (Fig. 6-15). Borehole recovery, when applicable, has the advantage of low installation and operating costs and minimal disturbance of the surface.

Heap leaching. In recent years the tremendous rise in the price of gold has resulted in the applications of heap leaching techniques to the recovery of gold from previously subgrade mine waste and mill tailings. Leaching techniques, using sulfuric acid or acid ferric sulfate solutions, sometimes with bacterial stimulation, have been used for processing low grade copper ores for many years and may be applied to gold (and silver) ores if a dilute cyanide solution is used as the solvent:

$$4 Au + 8 (CN)^- + O_2 + 2H_2O = 4 Au(CN)_2^- + 4 (OH)^-$$

Native gold is taken into solution as gold cyanide and recovered by adsorption on activated charcoal.

Figure 6-16 shows the essential components of the heap leaching process for gold

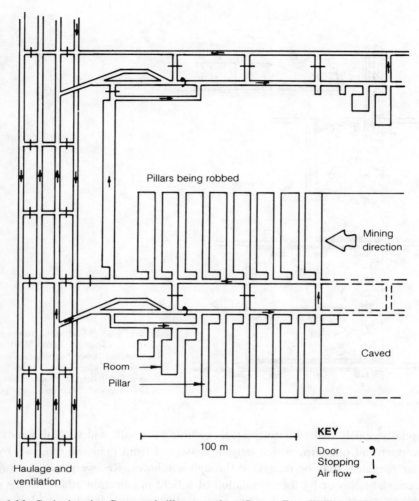

Figure 6.14. Coal mine plan. Room and pillar, retreating. (*Source:* From William H. Dennen and Bruce Moore, *Geology and Engineering*. Copyright © 1986 Wm. C. Brown Publishers, Dubuque, Iowa. All rights reserved. Reprinted by permission)

ores. One thousand to 50 million tonnes of ore is spread in a heap from 1 to 30 meters thick on a gently sloping impermeable pad and sprinkled with a weak cyanide solution for 7 to 90 days or longer depending on particle size and heap dimensions. The auriferous solution is drained to a "pregnant" pond and pumped through a series of three to five columns filled with activated charcoal. When the carbon in the lead column becomes fully loaded (200 to 400 ounces of gold per tonne), it is removed, the columns advanced, and a fresh column placed at the end. The gold is removed in solution from the loaded column by treatment with hot caustic solutions of sodium hydroxide plus some sodium cyanide and recovered as metal by electrolysis. The barren solution, after passage through the columns, is collected, made up to strength and adjusted for pH by the addition of NaOH or CaO, and pumped again through the sprayers thus closing the solution circuit.

Mining

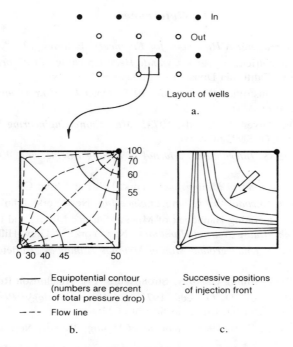

Figure 6.15. Arrangement of wells for in-place leaching. (*Source:* From William H. Dennen and Bruce Moore, *Geology and Engineering*. Copyright © 1986 Wm. C. Brown Publishers, Dubuque, Iowa. All rights reserved. Reprinted by permission)

Cyanide is very toxic, but the low concentrations used, recycling of the solution, careful handling of the reagents, and the fact that free cyanide is rapidly destroyed, does not accumulate in organisms, and can be metabolized in small quantities with no residual effects in humans have resulted in a very low incidence of accidents and adverse environmental effects in heap leaching applications.

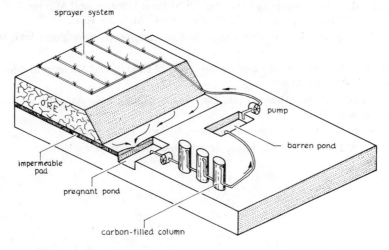

Figure 6.16. Principal components of a heap leaching operation.

References

Bray, R. N. 1979. *Dredging: A Handbook for Engineers.* Baltimore: E. Arnold.

Cole, K. A., and Kirkpatrick, A. 1983. *Cyanide Heap Leaching in California.* Sacramento: California Geology, California Division of Mines and Geology.

Crawford, J. T., and Hustrulid, W. A., eds. 1979. *Open Pit Mine Planning and Design.* Littleton, CO: SME of AIME.

Cummins, A. B., and Given, I. A., eds. 1973. *SME Mining Engineering Handbook,* Vols. 1 and 2. Littleton, CO: SME of AIME.

Farmer, I. W. ed. 1983. *International Journal of Mining Engineering.* London: Chapman and Hall.

Hoover, H. 1909. *Principles of Mining.* New York: Hill Publishing Co.

Hartman, H. L. 1987. *Introductory Mining Engineering.* New York: John Wiley & Sons.

Hoppe, R. ed. 1978. Operating Handbook of Mineral Surface Mining and Exploration, Vol. 2. *Engineering and Mining Journal Handbook.* New York: McGraw-Hill.

Hustrulid, W. A. 1982. *Underground Mining Methods Handbook.* Littleton, CO: SME of AIME.

Lacy, W. C., ed. 1983. *Mining Geology.* Stroudsburg, PA: Hutchinson Ross.

LeRoy, L. W., and LeRoy, D. O., eds. 1977. *Subsurface Geology: Petroleum, Mining, Construction,* 4th ed. Golden: Colorado School of Mines.

Lewis, R. S., and Clark, G. B. 1964. *Elements of Mining,* 3rd ed., New York: John Wiley & Sons.

MacDonald, E. H. 1983. *Alluvial Mining.* London: Chapman and Hall.

Martin, J. W., and others. 1982. *Surface Mining Equipment.* Golden, CO: Martin Consultants.

Merritt, P. C., ed. 1984. Book of flowsheets. *Engineering and Mining Journal Handbook.* New York: McGraw-Hill.

Peele, R., ed. 1941. *Mining Engineers Handbook,* 3d ed. New York: John Wiley & Sons.

Phleider, E. P., ed. 1972. *Surface Mining.* Littleton, CO: SME of AIME.

Sinclair, J. 1969. *Quarrying, Opencast and Alluvial Mining.* Amsterdam: Elsevier.

Sisselman, R. ed., 1978. Operating Handbook of Mineral Underground Mining, Vol. 3. *Engineering and Mining Journal Handbook.* New York: McGraw-Hill.

Stack, B. 1982. *Handbook of Mining and Tunneling Equipment.* New York: John Wiley & Sons.

Stewart, D. R., ed. 1981. *Design and Construction of Caving and Sublevel Stoping Mines.* Littleton, CO: SME of AIME.

Stout, K. S. 1980. *Mining Methods and Equipment. Mining Informational Services.* New York: McGraw-Hill.

Thomas, L. J. 1973. *An Introduction to Mining.* Sydney: Hicks Smith & Sons.

Young, G. J. 1932. *Elements of Mining,* 3d ed. New York: McGraw-Hill.

Chapter 7

Milling

PRINCIPLES

The goal of a milling operation is to provide one or more marketable products at a predetermined rate and quality by the efficient upgrading of run-of-mine ore. The products of a mill may be marketed directly, but many will require further processing by smelting or refining.

A number of devices are used in a mill to reduce the particle size of the input and free the individual mineral grains (comminution, liberation), provide uniformity of grain size (classification), remove ore minerals from gangue (separation) or impurity from a bulk commodity (beneficiation), and to provide for the disposal of the product and waste and the recovery of reagents (finishing). Generally, a number of different devices working in sequence and having appropriate side circuits are employed to accomplish these functions. The various devices work in concert with the output of one being the input of the next. These linked and interdependent units are arranged in series to improve product quality and in parallel to increase product quantity. The flow sheet or circuit for a mill thus has some of the characteristics of an electrical circuit since each device acts as a junction at which the sum of the outputs must equal the sum of the inputs. Individual mills face the need to handle ores from particular deposits and thus each mill will have a unique design in the kind and arrangement of its components. In general, however, the various processes used in milling may be categorized as shown in Table 7-1.

The profitable and effective operation of a mill requires that particular attention be given continuously to the optimum and balanced working of its various parts. Such tuning is accomplished by periodic sampling at critical points in the circuit, especially bottlenecks. In some instances automatic sampling and feedback controls may be appropriately installed.

PROCESSES AND DEVICES

Comminution and Classification

Efficient recovery and general mill operation requires that ore minerals be physically separated from particles of gangue or impurities from bulk materials. Ore and im-

Table 7-1.
Mineral Dressing Processes and Devices

Process	Property Exploited	Procedures or Devices Used
Comminution Reduction in particle size and increase in surface area	Tensile strength Compressive strength	Blasting Crushing: Jaw crusher, gyratory crusher, cone crusher, stamp, rolls, hammer mill, impact crusher
	Shear strength	Grinding: Disc grinder, rod mill, ball mill, stamp, buhr mill
Classification Sorting of particles by size and shape	Particle size Density and shape	Screening: Grizzly, screens, trommel, cyclone Settling pond, desliming tank cyclone, teeter column, hydraulic classifiers (rake, spiral drag, revolving rake)
Separation Separation and concentration of different phases	Appearance Density (separate minerals of same density by size and different density by species)	Cobbing Prospectors pan, shaking table, jigs (hand, movable sieve, fixed sieve), buddle, spiral, rheolaveur, sluice with riffles, vanner, heavy media
	Magnetic susceptibility	Magnetic separators
	Electrical conductivity	Electrostatic separators
	Wettability	Froth flotation, grease table
	Solubility and reactivity	Leaching, exchange reactions, adsorption
	Fluorescence	
Finishing Product, recovered reagents, and waste to final form and location	Density Solid/liquid Agglomeration	Settling of solids: Dewaterer, thickener, settling tank, clarifier Filter, drier Pelletizer

purity particle sizes thus determine the size to which material from the mine must be reduced before an efficient separation can be accomplished. The comminution is done by crushing and grinding, broadly distinguished by whether the material is broken by compression or shear. Figure 7-1 illustrates some of the commonly used devices.

Gyratory or jaw crushers with up to 2-meter gape followed successively by cone crushers in secondary comminution circuits and rod or ball mills for final pulverizing is a common sequence. Autogenous grinding in which finer particles are pulverized by coarser orefeed particles is more efficient than direct crushing and is used when

Milling

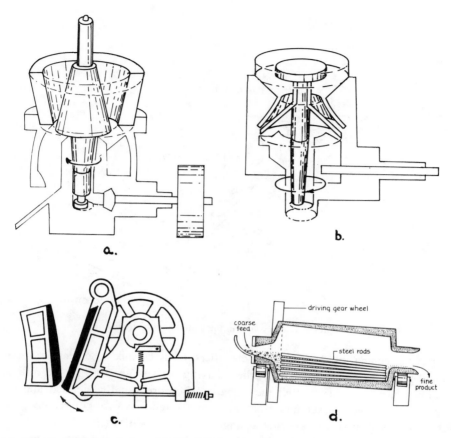

Figure 7.1. Comminution devices: (a) Gyratory crusher. (b) Cone crusher. (c) Jaw crusher. (d) Rod mill.

possible since one-half or more of the power consumed by a mill is in the comminution stage.

The output of a crusher or grinder will often vary more in size than makes for efficient operation of the separatory devices to follow. Screens or classifiers are then used to remove oversize material for recycling through the appropriate portion of the stage as illustrated by the circuit in Figure 7-2.

Free milling ores may often be beneficiated by screening of the output of the comminution stage because grains of ore and gangue minerals are not only disassociated but are probably of different sizes (Fig. 7-3). On the other hand, ore particles that are tightly locked will require comminution such that the coarsest particles produced are at least as small as the finest ore particles (Fig. 7-4).

The efficiency of separation of ore and gangue by one or another mechanical means is strongly dependent upon the grains being of uniform size. Larger low density grains may weigh the same as smaller high density grains and thus they may travel together rather than be separated. In milling this essential sizing is accomplished by

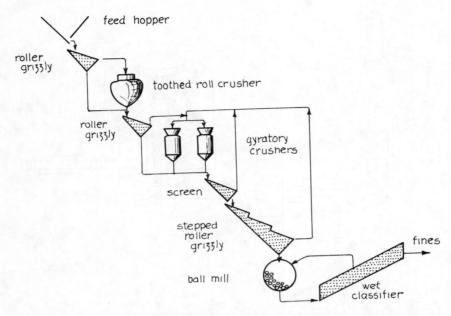

Figure 7.2. Recycling in the comminution circuit.

wet or dry screening, by devices called classifiers, or by cyclones. Screens are meshes, punched plates or parallel bars (grizzlies) having a fixed aperture so the mill feed will either pass over or through the surface. They are usually inclined and often vibrated to improve the screening action. Often screens with regularly decreasing aperture are used in sets to provide a range of sized products (Fig. 7-5).

Stirring or agitation of an incoherent mass of particles of the same size causes those of greater density to migrate downward; if size differs and density is the same, the smaller grains will migrate downward. This principle is exploited in classification (and separation by jigging) of ore-gangue mixtures. The operation is particularly effective in sorting smaller grains of higher density from larger and lighter fragments and is the basis for such devices as a rake classifier (Fig. 7-6).

Classification and separation by cyclones is widely used because such devices combine good partition ratios with large capacity and small physical size. Their op-

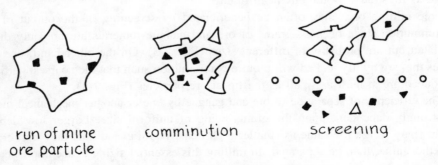

Figure 7.3. Size separation in a free-milling ore.

Milling

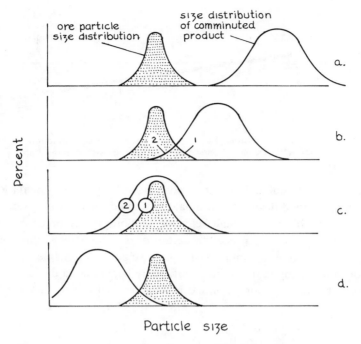

Figure 7.4. Unlocking. Key: (a) Comminution Stage 1. No ore particles unlocked. (b) Stage 2. Unlocking begins, some ore particles (1) are coarser than the comminution feed (2). (c) Stage 3. Unlocking continues, some ore particles (1) are still coarser than comminuted material (2). (d) Stage 4. Unlocking nearly complete.

eration involves the interaction of oppositely moving vortices of different density, which may consist of particles suspended in either air or water. Figure 7-7 shows the principle features of a cyclone.

Other means of classification are needed to separate those very tiny particles whose masses are so small with respect to their surface that surface phenomena dominate their physical behavior. Such particles, for example, encounter significant resistance in falling through air or water and separation can be made based on their rate of fall; this is elutriation.

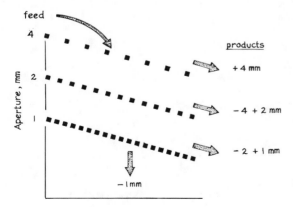

Figure 7.5. Product sizes from a set of screens.

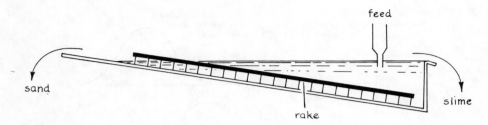

Figure 7.6. Principles of a rake classifier. Rake is drawn upslope, raised, and returned to the downslope position in a cycle of about 10 seconds. Coarser material is progressively moved upslope.

Because both the initial and operating costs of milling devices increase with size, it is incumbent in good design to eliminate unwanted material from the mill circuit at the earliest possible point. Hand selection (cobbing) may be economic in small operations, and contrasting physical properties of ore and gangue may sometimes be exploited in the comminution and classification stages. Soft or fine-grained minerals in a hard or lumpy gangue might be significantly beneficiated during comminution. For example, the tough, lumpy iron oxide gangue occurring with some bauxite ore

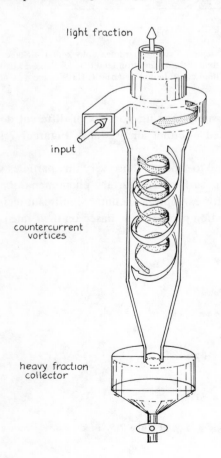

Figure 7.7. Dry cyclone classifier.

can be elimated from the mill by a grizzly and fine, soft cinnabar in a tough limestone will be enriched in the fines from crushing.

Separation

Following comminution and classification, ore flows to the separatory circuit in which one or more of the physical and chemical properties differing between ore and gangue minerals are exploited. The purpose of a mill is to remove unwanted impurities from a commodity or separate ore from gangue minerals in order to generate a marketable product. The heart of a mill is thus those separatory devices that perform this function. Some representative devices are described below.

The operation of a shaking table (Wilfley table; Fig. 7-8) involves the flow of a suspension of ore and gangue particles down a gently sloping surface that is regularly jerked parallel to its longer edge. Grains of the same size but different density have different inertia and, therefore, different resistance to accelerating forces. Thus particles of different density are separated into individual particle streams on the table and collected at different points along its end and lower edge.

A Humphrey spiral consists of a channel of semicircular cross-section spiralling downward (Fig. 7-9). The momentum of water-borne particles introduced at the top causes them to climb the curving wall to a height determined by the balance of gravity and lateral acceleration forces, a function of particle density. Particles of different density thus form into concentric streams at different radii. Properly placed holes tap off ore particles that are delivered via tubing to a collecting point.

Froth flotation is a much used procedure to separate sulfide ore minerals from their gangue. The principle is that hydrocarbon minerals or minerals with hydrocarbon coatings that can be selectively applied will stick to the surface of air bubbles in water. The practical application of this principle entails some preseparation treatment or conditioning by the addition of a small amount of liqid hydrocarbons to a suspension of particles to provide selective coating and the introduction of air and often a frothing agent to a suspensoid to cause frothing. Conditioned ore particles attach to bubble surfaces and the bubbles with attached ore particles rise to the top

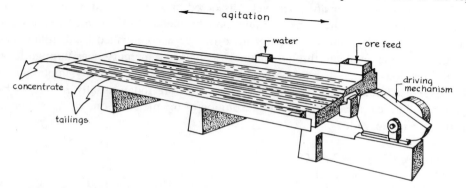

Figure 7.8. Shaking table (Wilfley Table). (*Source:* Redrawn after A. F. Taggart, *Elements of Ore Dressing.* New York: John Wiley & Sons, 1951)

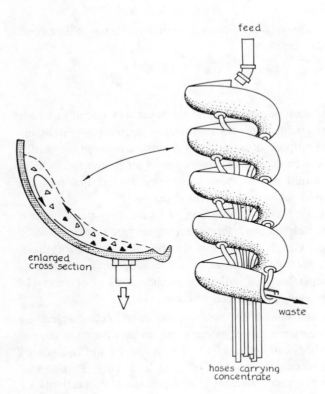

Figure 7.9. Humphrey spiral.

of the flotation tank or cell like a head on beer where they are skimmed off, whereas the gangue particles sink and are drawn off from the tank bottom (Fig. 7-10 illustrates the action). KEtX and $CH_3 \cdot C_6H_4 \cdot OH$ are organic coating and frothing agents, respectively, potassium ethyl xanthate and cresol in this case.

Special materials with special properties allow for the application of special procedures in mineral dressing. No attempt is made here to describe or even enumerate the many techniques that have been used for separations. A few examples to illustrate some better known instances would seem in order, however.

Native gold and silver are selectively dissolved or amalgamated by mercury and can be recovered from their finely ground ores by mercury treatment. Separation of the precious metals from the mercury may then be accomplished by distillation or by squeezing the amalgam through chamois leather; the mercury goes through leaving the precious metals behind. Amalgamation, although historically important, is no longer used commercially. The precious metals are today usually recovered by hydrometallurgical means involving the solution of the metal by one of several complexing agents to yield:

silver—$Ag(CN)_2^-$ or $AgS_2O_3^-$
gold—$Au(CN)_2^-$ or $AuCl_4^-$
platinum—Pt_6^{2-} or $PtCl_4^{2-}$
palladium—$PdCl_3^-$

Milling

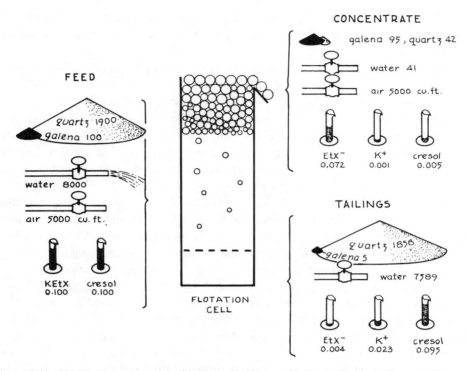

Figure 7.10. Idealized separations in a flotation cell. (*Source:* Redrawn after A. F. Taggart, *Elements of Ore Dressing*. New York: John Wiley & Sons, 1951)

Recovery of silver from $AgS_2O_3^-$ is by regeneration with $NH_4S_2O_3$ and all others by roasting. The development of the technique of heap leaching in which small gold contents are dissoved from old mine and mill wastes by the percolation of a weak sodium cyanide solution through them has coincided with the dramatic rise in gold prices in recent years. The gold may be recovered by adsorption on activated charcoal followed by roasting.

Diamonds, unlike most other minerals, will stick to grease. This makes diamond rings hard to keep clean, but also provides a means for diamond recovery by passing ore over a greased surface where they stick. The diamonds are easily separated from the grease by gentle heating and screening.

Copper ions in solution will displace iron atoms from metallic iron, thus gradually transforming a piece of iron to a copper powder. Copper is recovered from mining wastes by bringing the weak, copper-bearing solutions that percolate through a dump into contact with scrap iron. Efficiency is improved by seeding the water with particular bacteria that accelerate the solution of copper from its minerals.

Solvent exchange processing, variously called SIX (solvent ion exchange), liquid-liquid ion exchange, or LIX (liquid ion exchange; trademark of General Mills Chemical Company) has become the preferred means of processing copper sulfide and uranium ores in the past two decades. Ores are leached with sulfuric acid, sometimes in combination with ferric chloride, the metal extracted from the leachate by an organic reagent dissolved in a kerosene-type diluent, stripped from the loaded organic phase into an aqueous phase, and the metal recovered by electrolysis, precipitation, or crystallization.

Separation principles. Improved product quality may be obtained by installing separatory circuits in series. If the efficiency of single stage devices is insufficient to obtain optimum quality, a number may be connected so that each successive device operates on the upgraded output* of the preceding unit. This, however, is a system of diminishing returns since upgrading becomes successively more difficult and some losses occur at each stage. Under these circumstances a point will sooner or later be reached at which the purchase or maintenance of another stage is not warranted by the increase in product quality.

The operation of a separatory device more or less cleanly divides an input mixture into two or more streams. The action may be thought of as simultaneously dividing grains of mineral A and B (or A, B, C . . . N) between two outflow channels in changed proportions. For example, an input mixture having the proportions of 5A to 16B might be divided into two streams in which the respective ratios were 1A to 14B and 4A to 2B. Ideal separations yielding 100% of a given component free from contamination may be approached in very efficient separatory devices, but are never attained. The reasons are various and include the presence of two components in a single grain due to incomplete unlocking during comminution, the physical sweeping of some grains of one component along with the flow of another, too wide a range of grain sizes in the feed, and inefficient surface preparation in devices involving wetting, surface charge, or chemical reactivity.

Figure 7-11, in which the proportions of two materials passing through a separatory unit are indicated by the areas of the blocks, shows diagrammatically the principles of a separatory stage. The input is a mixture of one part A (black particles) and three parts B. The input grade is thus 25% A. In separation, assume A is divided between outputs 1 and 2 in the proportions of two to one or 0.67 to 0.33. Simultaneously, assume 0.2 of B is channeled to output 1 and 0.8 to output 2.

If the input consists of 100 tpd (tonnes of ore per day), 25 of A and 75 of B, the flow through output 1 will consist of:

$$25(.67) + 75(.2) = 31.75 \text{ tpd whose grade is}$$
$$25(.67)/31.75 \times 100 = 52.75\% \text{ A}.$$

Similarly, output 2 will be

$$25(.33) + 75(.8) = 68.25 \text{ tpd with a grade of}$$
$$25(.33)/68.25 \times 100 = 12.1\%.$$

The grade of A has been improved in output 1 or, alternately, the purity of B has

---

*Mill usage terms both inputs and upgraded outputs as "heads" and outputs for discard as "tails" or "tailings." This terminology is avoided here.

Milling 145

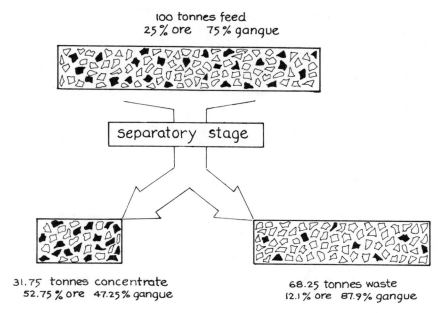

Figure 7.11. Principle of a separatory stage.

been improved in output 2. Note, however, the loss of component A through output 2.

It is often the case that a single stage is inadequate to achieve the separation required and units are connected in series, the upgraded output of one being the input to the next. Figure 7-12 shows diagrammatically how the partition of ore and gangue occurs in serial stages. The numbers may be taken as tonnes per day, the first being ore and the second gangue, initial input in this case being taken as 100 tpd of ore having a grade of 10%. It should be noted in Figure 7-12 that whereas the grade of the product is improved by sequential stages, the tonnes of product is continuously reduced. This phenomenon of increased grade being coupled with lessened recovered quantity is illustrated in Figure 7-13 (data from Fig. 1-12). Such curves drawn from test data from particular separatory devices and ores provide insight regarding recovery factors, losses in tonnage for gains in grade, and allow comparison of competing systems when choosing mill machinery and designing circuits. A point to note is that it is the recovered ore that is the basis for true tonnage calculations when determining the total value of a mine.

Marketing contracts are usually at a fixed price per tonne* of agreed grade coupled with a scheme of bonuses for higher and penalties for lower grades. Because the action of sequential separatory devices in a mill improves output grade while reducing tonnage, there must be some optimum number of stages. As an example, assume an agreement of $200 per tonne of 70% grade, bonus of $5 per tonne for each percentage point above 70%, and a penalty of $6 per tonne for each point

---

*Different commodities are sold in different units; see Appendix III. Tonne is used here as representative.

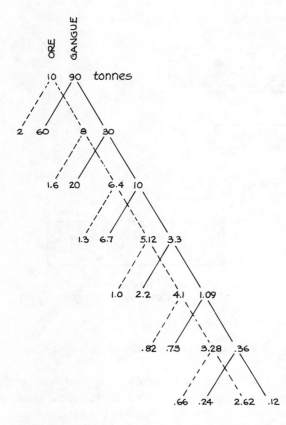

Figure 7.12. Partition of ore and gangue minerals in a series of separatory stages. Ore mineral partition 4:1, gangue mineral partition 2:1.

below. Table 7-2 is derived from Figures 7-12 and 7-13, and it is obvious that maximum profitability is attained after four stages of separation.

Because each stage represents a fixed change in tonnage and grade, it may sometimes be more profitable to overshoot the market specifications and then reduce the product grade by blending the output with some discards. This is particularly pertinent if there is no bonus schedule.

In the present illustration, a bonus of $45 per tonne is being awarded for an output of 5.19 tonnes per day, whereas 3.23 tonnes at a grade of 31.6% are being discarded at the same stage (Fig. 7-12). Blending of 1.2 tonnes of this nominal discard to yield a product of 70% grade gives a product whose value is $1278 per day, a small but significant improvement in income.

$$\frac{.79(5.19) + .316x}{5.19 + x} = .70, \quad x = 1.2 \text{ tonnes}$$

The initial quality of mined material and efficiency of separatory devices are obviously of great importance in determining the profitability of a mining operation. As a further illustration of this, consider the following circuit intended to reduce the impurity content of a commodity to below 0.10% from an initial content of 1.00%. The partition efficiency for each stage in the circuit is 75%. Numbers in parentheses

Milling

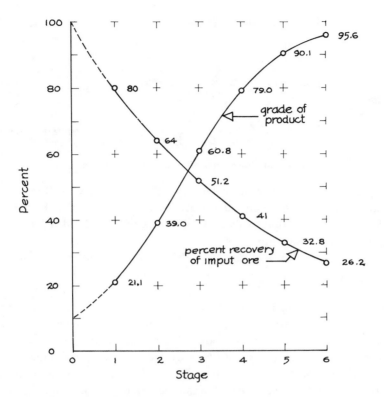

Figure 7.13. Change in product grade and tonnage in successive separatory stages.

in Figure 7-14 are percent impurity. Only a bit more than half of the potential commodity is being recovered and large amounts of waste must be disposed of. The obvious challenge is for the geologist to locate a deposit of greater purity or the mineral dresser to improve the efficiency of separation.

Many ores consist of more than one mineral of value and it is often of economic importance to recover several coproducts from milling. As an example of a simple operation, consider an ore containing 10% galena, 20% barite, and 70% calcite from

Table 7-2.
Optimization of Separation

Stage	Product Grade	Market Grade (70%) Minus Grade of Product at Stage, %	$ Differential	$ Value Per Tonne	Tonnes Produced	Product Value Per Day, $
1	21.1	−48.9	−293.40	—	38	—
2	39.0	−31.0	−186.00	14.00	16.4	229.60
3	60.8	−9.2	−55.20	144.80	8.42	1219.22
4	79.0	+9.0	+45.00	245.00	5.19	1270.55
5	90.1	+20.1	+100.50	300.50	3.64	1093.82
6	95.6	+25.6	+328.00	328.00	2.74	898.72

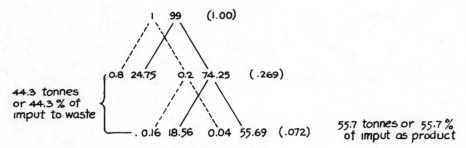

Figure 7.14. Beneficiation by serial separatory stages.

which it is desired to recover the galena and the barite. Two different separatory devices are assumed to be required for efficient separation, which, when used in series, yield daily production totals of 11.4 tonnes of galena, 14.8 tonnes of barite, and 73.8 tonnes of waste (Fig. 7-15).

Finishing and Transportation

The final steps in milling are the preparation of the product for shipping and the disposal or reuse of the various other materials that have been separated. The product must be dried and bagged or bulk-stored awaiting shipment, reagents whose value exceeds their cost of recovery should be recycled, and the often large and environmentally degrading wastes must be disposed of. The latter is a particularly serious problem because of the sheer bulk of waste generated in most mining and milling activity. However, waste disposal is not usually considered as a component of finishing, and so it is discussed in the next chapter.

The transportation system of a mine and mill might be described as stope to ore pass to level to shaft crusher to ore pocket to measuring pocket to skip via shaft to surface and by surface transfer to surge bin to primary crusher to waste disposal and secondary crusher to weighing/sampling station to fine ore bins to wet grinding plant to ore treatment plant to waste disposal and weigh/sampling station to dispatch of concentrates by surface or water transport. Such a system involves the use of gravity chutes, underground and surface truck or rail haulage, conveyer belts, shaker chutes, slurry lines, aerial tramways, and such as transporting devices and many transfer points, surge bins, and measuring stations. It is obvious that every attempt should be made to minimize transport distances and number of transfer points and to utilize the most cost-effective transportation means.

The most common transport device used in the minerals processing industry is a belt conveyor—a rubber-coated fabric band that passes around two cylindrical pulleys, one driven, and is supported by idlers at regular intervals. Such conveyors may be used for either horizontal or inclined transport of dry material, provide high capacity with low power requirements, and deliver uniform volumes at high speeds.

Slurry pipelines can be economically effective in many situations where topography is favorable and water readily available. Special pumps are used to force finely ground solids suspended in water through pipes of relatively small diameter, and

Milling

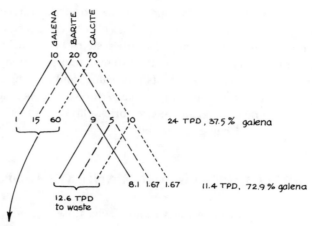

Because the initial device is not providing an efficient separation of barite and calcite, this output is used as feed to a different kind of unit delivering barite and calcite in the ratio of 90:20 to one outlet

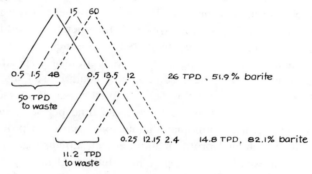

Figure 7.15. Use of different separatory sequences. Since the initial device is not providing an efficient separation of barite (input 20 tpd) and calcite (70 tpd), this output is used as feed to a different kind of unit delivering 50% of the galena, 80% of the calcite, and 10% of the barite to one output.

dewatering facilities are installed at the output end. Examples of slurry lines handling concentrates between mills and shipping or processing points are:

Material	Location	Pipe diameter	Pipeline length
magnetic concentrate	Savage River Mines, Tasmania	9 in	53 mi
hematite concentrate	Samarco Mineracao, Brazil	20	246
copper concentrate	Bouganville Copper Ltd., Bouganville	6	20
copper concentrate	Ertsberg Project, Indonesia	4 1/4	69
copper concentrate	Pinto Valley Mill, Arizona	3 1/2	10.7
phosphate concentrate	Tapira, Brazil	9	74

Transport costs within the mine and mill and to markets have been estimated to make up one-quarter to one-third of the final cost to the consumer. Producers with the same ore and equally efficient mine-mill systems but at different distances from a given market may have significantly different transportation charges. Thus ore acquires a place value in addition to its unit value, and the impact of place value is particularly evident in the marketing strategy for such bulk materials of low unit value as sand and gravel, cement rock, or brick clays. (Further aspects of transportation are discussed in Chapter 11.)

PRINCIPLES OF MINE AND MILL DESIGN

The integrated arrangement of the various devices used in ore processing will obviously take different forms depending on the many factors that influence efficient separatory operations. Although no single arrangement or circuit may be taken as typical, a few examples of illustrating different formats are given in Figure 7-16.

The design of mills and mines is dictated by both executive and engineering decisions; the former dealing with the place of the operations within the larger frameworks that often include activities integrated from mines and mills through smelters, fabricating plants, and marketing systems. Selection of a specific deposit lifetime, probably made in view of perceived economic conditions, fixes the production rate and allows calculation of operating costs, capital expenditure and debt service, taxation, and other costs to be made for a fixed time period. Alternately, executive concerns or engineering factors may dictate some particular production rate perhaps based on a desired annual profit or requirements downstream in integrated operations, which then establishes the operating lifetime. Usually some compromise between a predetermined lifetime or one dictated by the production rate will be established.

The design capacity of the mine and mill is thus fixed by decisions regarding either their predetermined lifetime or their production rate. To recover the ore within a fixed lifetime, the plant must be able to handle the total deposit at a determinable rate and mining and milling design must proceed from input (heads) to output (product and tailings) allowing for adequate annual capacity. A plant designed to provide a product at a predetermined rate, on the other hand, must be designed from output to input.

Capital costs of new mines mills may be expected to increase because of increased costs of labor and materials, needs for more automation, and greater material handling capacity. Careful design and utilization of cost-effective construction materials and practices are thus essential.

Mill Circuits

Head to tail design for mills. Head to tail design for a mill is done by making the capacity of the input equipment, usually the section for crushing and classification, adequate to handle mine run inputs plus any recycled materials. The capacity of the first stage in the following separatory circuit must be able to handle this volume, but the capacity of downstream devices is reduced by the amount of waste removed by its upstream predecessor.

As an example of flow sheet calculations for a mill, consider a simple circuit in which oversize material is recycled through a primary crusher and the following separation circuit consists of two devices in series (Fig. 7-17). The important criterion is that the sum of all outputs at any point in the circuit must equal the sum of all inputs. The capacity of the crusher must be equal to the sum of mine-run ore plus recycled oversize; capacity = 80 + 20 = 100 tpd (tonnes per day). The grinding efficiency = 100 − 20/100 = 0.8.

The capacity (duty) of separatory stage 1 must equal the crusher output of 80 tpd. The separation ratio for this stage is 20/20 + 60 = 1/4. The capacity of separatory stage 2 must accommodate the product output of stage 1, 80 tpd × 1/4 = 20 tpd. The separation ratio for this stage is also 1/4 and yields 5 tpd of product. (Daily recovery is 5/80 × 100 = 6.25%.) Product storage must accommodate 5 tpd and waste disposal 75 tpd. 75 + 5 = 80, the output of the crusher.

Figure 7-16c shows the amount of material passing through the various parts of a real mill and indicates diagrammatically the capacities of the various devices in use.

Tail to head design. Tail to head design proceeds upstream through the mill circuit beginning with the final output product rate and assigning capacities to compensate for waste removal and recycling until the initial input is found.

Assume that a final product of one tonne of concentrate is required daily from an integrated mine-mill operation in order to maintain the cash flow, defray all costs and provide a reasonable profit. Further, assume that the in-place ore grade is 1.0%, 75% recovery is possible, and no stoppages are encountered. The daily output of 1 tonne will require the mine to produce and mill to process 133 tonnes of ore per day:

$$x (0.75 \times 0.01) = 1, x = 133$$

This capacity, in turn, dictates such features as the number of working faces in the mine, the size of the mine transport system, scale of milling operations, size of the waste disposal areas, and indeed all aspects of the operation

Note that in tail to head design it is the product value that is controlling, so changes in market price on one hand or ore grade on the other may force modifications in the mine-mill operation, increase profitability, or cause the operation to fail. Insofar as possible, anticipation of such contingencies should be incorporated in physical planning.

In all designs for mills it must be recognized that the various devices used have fixed capacities and thus some will be underused and some stage or unit will be the bottleneck that limits production rate. Overcapacity of a device may be accommodated by operating it part-time and providing for input and output storage. Adequate provision must always be made for unforeseen stoppages, safety of workers and

152 Extraction and Milling

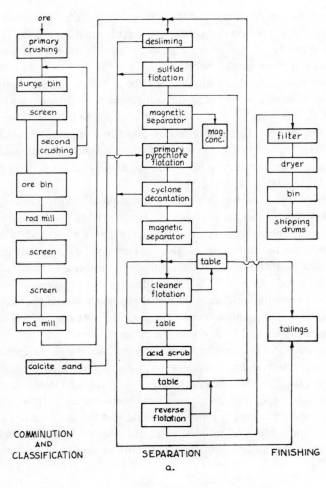

a.

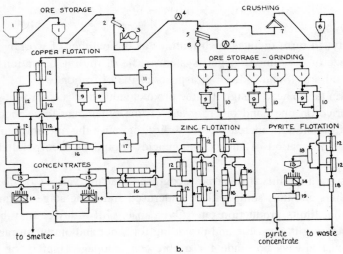

b.

Milling

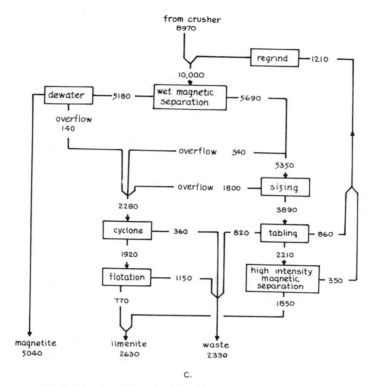

Figure 7.16. Examples of mill circuits. (a) Concentrator flowsheet, St. Lawrence Columbium & Metals Corp., Oka, Quebec. (*Source:* From C. Carbonnear and J. H. Foucher, Columbium Raw Material, in *Milling Methods in the Americas*. Copyright © 1964, Gordon & Breach Science Publishers Inc, New York. N. Arbiter ed); (b) Concentrator flow sheet for copper-zinc-iron sulfides, Normetal Corp. Abitibi Co., Quebec. (*Source:* Reprinted with permission of the Canadian Institute of Mining and Metallurgy from *The Milling of Canadian Ores*. 6th Mining and Metallurgical Congress, 1957. Key: (1) storage bin. (2) grizzly. (3) jaw crusher. (4) magnet. (5) screen. (6) surge bin. (7) crusher. (8) weightometer. (9) ball mill. (10) spiral classifier. (11) aerator. (12) pneumatic flotation cell. (13) thickener. (14) settling tank. (15) filter. (16) mechanical flotation cell. (17) Conditioner. (18) cell cleaner. (19) dryer. (**c**) Tonnages in various parts of a mill. (*Source:* From B. Hower, The Macintyre Concentrator, in *Milling Methods in the Americas*. Copyright © 1964, Gordon & Breach Science Publishers Inc., N. Arbiter ed. New York)

equipment, and tuning or adjustment of the circuits components. Changes in milling capacity may be made by installation of parallel circuits and the addition of devices in series will serve to upgrade the product. Surge storage at critical points guards against shutdown for lack of ore and grizzlies and magnets guard against oversize rocks and tramp iron.

The coupling of milling operations with mine production is of considerable importance. Milling is typically carried on continuously, whereas mining tends to be conducted as cyclic sequence of operations involving such steps as drilling, blasting, loading (mucking), and hoisting. To provide a smooth flow of ore to the mill it is common to provide some temporary ore storage both at the mine and at the mill.

A mine in an integrated mine-mill partnership should produce a predetermined quantity and grade of ore. For variable ores, this may require the simultaneous ex-

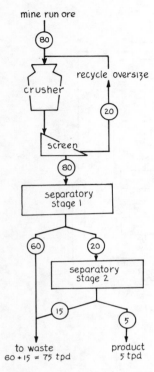

Figure 7.17. Determination of capacity (see text for details).

cavation of ore of different amounts and quality from different parts of the mine, which is then blended, usually by delivery to the mine head on a schedule that provides the necessary increments of ore grades and amounts.

WASTE DISPOSAL AND RECLAMATION

Mining and milling may generate very large volumes of waste that require disposal in a manner consistent with the requirements of the following acts (as well as various other federal, state, and local statutes):

1. Refuse Act of 1899: unlawful to discharge or deposit wastes into navigable waters or their tributaries.
2. National Environmental Policy Act of 1969: defines a national policy that will encourage harmony with the physical environment and promote efforts to prevent environmental degradation.
3. Water Quality Improvement Act of 1970: water pollution control law setting water quality standards.
4. Resource Conservation and Recovery Act of 1976: statutory controls of hazardous wastes.
5. Surface Mining Control and Reclamation Act of 1977: controls of environmental effects of strip mining including prohibition of past practices that have

led to environmental degradation and required postmining reclamation of the surface.

Table 7-3 suggests one aspect of the waste disposal problem—the great volume that is produced. The increase in porosity attendant on comminution generates a 30 to 50% increase in effective volume over the same material in place, so even with (usually) small amounts removed as product and return of waste to underground or open cast workings, a very significant volume remains.

The second aspect of the disposal problem is the common presence in the waste material of undesirable and occasionally toxic elements, either from unrecovered ore minerals, the gangue, or mill reagents. The easy access of water and air to waste fragments may result in leaching with consequent pollution of surface and ground waters. A third important aspect of waste disposal is the need to protect surface drainage from siltation resulting from erosion of fine-grained waste material. Less critical but locally important concerns have to do with dust, odor, noise, and esthetic aspects of disposal sites.

The means of waste disposal depend upon the nature of the material and local geologic conditions. Heaps of waste (culm dumps, slag piles, gob piles, etc.) are a common and satisfactory means of disposal of nonreactive and coarsely granular wastes from the operational point of view but are admittedly unsightly. If erodable fine material is present, silt ponds must be constructed below the waste disposal area to trap water-borne debris. Very fine material must be held in slime ponds in order to allow the suspension of impalpable solids in water to settle and the water to drain away or evaporate.

Noxious or toxic chemical components in waste must be removed or neutralized either before the waste is discarded or after it is in place. This may be done by chemical reaction such as lime treatment to neutralize acidic waters, precipitation

Table 7-3.
Wastes Generated by the Mineral and Fossil Fuel Industries, 1975

Industry	Mine Waste	Mill Tailings	Washing Plant Rejects	Slag	Processing Plant Wastes	Total (Thousands of Tons)
Copper	680,400	266,800		5,200		952,400
Iron and steel	257,900	154,600		26,000	2,200	440,700
Bituminous coal	12,800		107,100			119,900
Phosphate rock	216,000		137,300	4,000	50,000	407,300
Lead-zinc	5,200	17,400	1,000			23,600
Aluminum					14,700	14,700
Anthracite coal			2,000			2,000
Coal ash					67,800	67,800
Other[1]					285,900	285,900
TOTALS	1,172,300	438,800	247,400	35,200	420,600	2,314,300

[1]Estimated waste generated by remaining mineral mining and processing industries.
Source: C. Rampacek, 1982. Appendix H in *Geological Aspects of Industrial Waste Disposal*. (Washington: National Academy Press, 1982).

with appropriate reagents, or adsorption on clay, zeolites, or activated charcoal. When possible, such wastes should be returned underground if there is no possibility of their contaminating groundwater supplies.

Mine and mill wastes are materials that have been discarded because they were not economically viable at the time of their production. Costs of their mining and comminution have, however, been paid and the material is readily accessible. Increases in the value of commodities through time and improvements in recovery technology can sometimes make the waste of earlier operations the orebody of today and re-examination of the possible utilization of mining and milling wastes is an ongoing activity.

SUMMARY

The economic viability of a deposit depends not ony on its in-place value but also on the organization and efficiency of its extraction and processing. Workers and machines must be employed in optimum numbers and combinations to attain a predetermined production rate. Known or reasonably estimated parameters are total tonnage and grade of the deposit, machinery and labor costs of mining and milling, percent recovery, unit value of the commodity in the marketplace, transportation costs, and the cost of money.

The geologic considerations of importance in the evaluation of a mineral property and the means whereby the data might be acquired are described in Part I. Further parameters affecting a property's value are discussed in this part and recapitulated in outline form below:

1. Geologic considerations lead to determination of the in-place tonnage and grade of a deposit. These values multiplied by a prediction of market value at the time of sale* yield an initial in-place value.
2. Tonnage × Grade × Value per Tonne = Total Value. Initial tonnage must be debited by ore left in place during extraction and nonrecovery during milling.

$$(\text{Tonnage} - \text{Mining Losses}) \times \text{Grade} = \text{Mined Value}$$

and

$$(\text{Tonnage} - \text{Mining Losses}) \times \text{Mill Recovery} \times \text{Grade} = \text{Recoverable Value}$$

3. Economic assessment must be made on recoverable value.

References

Arbiter, N., ed. 1964. *Milling Methods in the Americas*. New York: Gordon and Breach.

Denev, S. I., Stoitsova, R. V., and Boteva, A. D. 1984. *Selective Processing and Flotation of Ore Minerals*. Athens: Theophrastus Publications S.A.

---

*Usual practice is to evaluate in terms of present dollars and extrapolate from market trends to future value.

Fuerstenau, D. W., ed. 1979. *Flotation*. Littleton CO: SME of AIME.
———, and Miller, J. D. 1985. *Chemistry of Flotation*. Littleton, CO: SME of AIME.
Gaudin, A. M. 1957. *Flotation*. New York: McGraw-Hill Book.
Great Britain Department of Scientific and Industrial Research. 1958. *Crushing and Grinding; a Bibliography*. London: H. M. Stationery Office.
Kelly, E. G. and Spottiswood, D. J. 1982. *Introduction to Mineral Processing*. New York: John Wiley & Sons.
Lowrison, G. C. 1979. *Crushing and Grinding*. London: Butterworth.
McQuiston, F. W., Jr., and Shoemaker, R. S. 1978. *Primary Crushing Plant Design*. Littleton, CO: SME of AIME.
Mular, A. L., and Bhappu, R. B., eds. *Mineral Processing and Plant Design*, 2d ed. CO: SME of AIME, Littleton.
———, and Jergensen, G. V., II 1982. *Design and Installation of Comminution Circuits*. Littleton, CO: SME of AIME.
Newton, J. 1959. *Extractive Metallurgy*. New York: John Wiley & Sons.
Pryor, E. J. 1965. *Mineral Processing*, 3d ed. New York: Elsevier.
Richards, R. H., and Locke, C. E. 1940. *Textbook of Ore Dressing*. New York: McGraw-Hill.
Roberts, A., ed. 1963. *Mineral Processing*. London: Pergamon Press.
Taggart, A. F. 1951. *Elements of Ore Dressing*. New York: John Wiley & Sons.
Weiss, N. L., ed., 1985. *SME Mineral Processing Handbook*. Littleton, CO: SME of AIME.
White, L., ed. 1980. *E/MJ Second Operating Handbook of Mineral Processing*. New York: McGraw-Hill.
Wills, B. A. 1981. *Mineral Processing Technology: An Introduction to the Practical Aspects of Ore Treatment and Recovery*, 2nd ed. New York: Pergamon Press.

Part III

Economics, Regulation, and Trade

Chapter 8
Economic Considerations

INTRODUCTION

The economics of a mining venture are embedded within the larger framework of national and international economics. In the scope of this book, the larger picture cannot be explored, but it is important to understand those special features peculiar to or making significant impact on the economic aspects of the mineral world because they impinge at every point on the technical planning and development of mines and mills.

The technical features that are used to define the value of a mineral deposit are described in previous chapters. Given this in-place value, it is next necessary to consider all of the costs to place the commodity in the marketplace. Remember, the deposit is an ore only if its in-place value is sufficient to cover all of the attendant costs from mine to market plus a margin of profit.

A significant constraint on the economics of mineral production is the fact that an orebody is a wasting asset, exhausted by exploitation and irreplaceable. An orebody has no intrinsic value when undiscovered in the ground; discovery accords value upon anticipated future profitability and true valuation must account for all exploration and exploitation costs adjusted for time. The real value of a mineral deposit may thus be defined as the total profit that may be gained by its working.

The lifetime of a mining operation is a function of the size of the orebody and the rate of its extraction. The former is geologically fixed and the latter chosen on engineering and economic grounds to yield maximum profitability. Anticipated lifetimes will usually lie in the range of 10 to 30 years, although lifetimes of many mines have been much longer than this. Some, such as the salt times of Austria and the copper mines of Germany, have been worked in excess of 700 years without interruption; travertine has been quarried in Italy since Roman times; and copper mining on Cyprus was well established by 1500 B.C. and has continued to modern times. Mining in some districts—Cornwall, England, for example—has continued since the Neolithic era. However, it is romance and not business to hope for new discoveries, new technologies, or new economic factors to extend the life or increase the profitability of a deposit. Properly and conservatively planned, a known orebody is to be exhausted, the plant fully depreciated, operating costs and profits paid, and a sum sufficient to acquire an equivalent deposit accumulated during its anticipated operating lifetime. It is common to use 10 years as the basis for initial engineering and economic considerations. This is a reasonable period for forward planning of technical activities and discount factors at usual interest rates have fallen well below 0.5 over this time.

SUPPLY AND DEMAND

The classic supply-demand curves of economics (Fig. 8-1) are basic to an understanding of the nature of the marketplace. Even though the actual operation of the marketplace is beset by restrictions and controls, the curves describe fundamental aspects of any economy, even one without free markets.

The equilibrium between price and available quantity is shown by the point of intersection of the two curves. It is this point that is sought by changes in supply and demand consequent on changes in price and quantity. If a commodity is overpriced as at level a, consumers will buy less and overstocked sellers will lower prices. Reciprocally, prices at level b create a sellers market in which competitive buyers will bid up the price. Identical relations exist when the available quantity of a commodity is considered with level c one of competitive bidding for scarce materials and level d a buyers market in an overstocked situation. These supply and demand pressures will, in a free market, drive price and quantity to the point of intersection of the curves.

Supply-demand curves are obviously idealized representations of the workings of the marketplace and should be used to understand it, not predict it. Curve shapes and orientations may vary from essentially horizontal straight lines—no change in price regardless of quantity available or demand regardless of availability—to vertical in which there is no change in the amount of demand or supply regardless of price. Under circumstances in which there is a general increase in supply under a condition of constant demand, the supply curve moves to the right and the price level of the curves' intersection falls, whereas a decrease in supply drives the price up in a like manner.

Trading in commodities takes place because of the pressures to establish a more or less perfect balance of supply and demand, the economic verity that operates even when modified by national politics, monopolism, and reciprocal trade arrangements.

Producers and users of mineral commodities are necessarily concerned with the price and availablility of materials at the present time and into the future. This prediction may be done in a qualitative but none-the-less useful way by considering the impact of various possible future events on the present-day equilibrium point. As examples, the following are reasonable events that might affect mineral economics:

Decrease in quantity
    The orebody of a principal producer becomes exhausted.
    The country in which a principal producer is located declares an embargo on shipment to a using country.

Economic Considerations

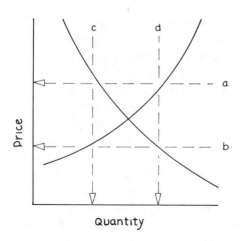

Figure 8.1. Supply and demand versus price and quantity.

Industrywide strikes curtail production.

War places the producing country in the enemy camp.

Political instability reduces capital investment in the producing facility.

Increase in quantity

A newly discovered deposit is placed in production.

Higher prices encourage the opening or reopening of known marginal deposits.

Technological improvement in milling allows greater recovery.

Decrease in demand

Cheaper substitutes for the commodity are found.

Health concerns reduce the use of the product.

Procedures for recycling are found.

Prices are fixed at a high level by cartel or government action.

Increase in demand

Increased production of manufactured goods using the raw material.

Price is supported by government subsidy.

Government purchases are made for a stockpile.

Both the supply and demand for many commodities are subject to the artificial control of cartels and governments. Production and price fixing by cartels or monopolies for oil, tin, cobalt, bauxite, iron ore, and diamonds are, or have recently been, established features of the trade in these commodities. Governmental economic control through subsidy, import tariff, stockpiling, and tax incentives are mechanisms used to maintain adequate supply levels at acceptable prices for domestic producers, whereas restrictive regulations, excessive taxation, sequestration of land from industrial use, and export embargoes reduce domestic activity. The cost of money as interest on borrowed capital is strongly dependent on the state of national economy and can markedly affect mineral development.

Technological changes and community attitudes must also be considered as affecting supply and demand. Substitution, recycling, conservation, innovation, and obsolescence all have, and will, take place. Prestressed concrete is an excellent choice as a replacement for steel in bridge girders, magnesium makes lighter canoes than aluminum, thorium-impregnated gas mantles went out with electric lighting, scarce tin supplies during World War II were supplemented by recycling tubes and tinned cans, lead paint is not used for health reasons and titania pigments have replaced it, the nuclear age has made uranium and zirconium into important metals, and synthetic pigments, abrasives, and gems have supplanted their natural counterparts. U.S. imports of industrial diamonds, for example, have been halved in the past 20 years and replaced by synthetic stones. What will recently announced electrically conducting plastics do to the market for copper?

Primary producers and consumers commonly negotiate long-term agreements with respect to sale price and quantity, usually with clauses regarding regular renegotiation. Such agreements are mutually beneficial because assurance of a future relationship is made. Neither seller nor buyer is necessarily restricted by such agreements to a single trading partner, however, and both typically have several contracts in force. Excess production (or needs) not covered by agreements are absorbed in the commodities spot market where daily trading is carried on. The principal mineral commodities markets are in New York (COMEX) and London (London Metal Exchange).

Large deposits of valuable commodities are far from uniformly distributed over the Earth; rather, they tend to be concentrated in a few localities so the supply for some materials may be dominated by one or a few major mines. A similar situation may exist for the principal users of these commodities, and a small community of suppliers and users arises. Such a community is particularly vulnerable to dislocations caused by the exhaustion or discovery of deposits on the one hand and changes in demand for whatever reason on the other. The free world community of supplier-user nations for nonfuel mineral commodities is dominated today by the United States, Canada, Great Britain, Japan, Australia, and the Republic of South Africa.

The consequences of changes in supply and demand in such a small community may be seen by consideration of Figure 8-2. In a free market (Fig. 8-2a), suppliers A and B compete with each other for sales to users 1, 2, and 3. If the selling price of A is less than that of B (Fig. 8-2b), A will capture the greater share of the market. Conversely, if 1 is willing to pay a premium price, the distribution pattern will shift to the detriment of 2 and 3 who are then faced with the choice of paying more, substitution, or conservation (Fig. 8-2c).

An increase in market value (Fig. 8-2d) will encourage suppliers to increase their production leading toward oversupply and reduction of market price. Conversely, an increase in selling price by a cartel of suppliers (Fig. 8-2e) will decrease demand and encourage exploration for uncontrolled sources, substitution, and conservation.

Imperfect supply-demand situations will arise (Fig. 8-2f) because of reciprocal agreements between trading partners, say A-2. The arrangements of B-2 and B-3 in the example are unchanged, but 1 has been cut off from A, and a new B-1 arrangement may next be established that encourages expanded production by B to meet the additional demand of 1. Both total production and consumption have thereby been increased.

Economic Considerations

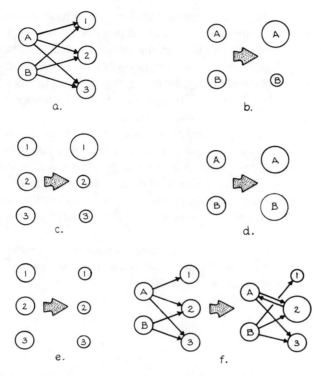

Figure 8.2. Supply and demand changes in a small community (see text for details).

Should a new supplier enter the market, supply will obviously rise and the excess supply will depress prices on the world market. Reciprocally, the loss of supply will generate a rise in price.

CAPITAL FORMATION AND CASH FLOW

The financing of mining ventures requires a delicate balancing of risks and rewards. Risk, by definition the probability that a desired outcome will not be achieved, is very high at the earliest stages of exploration and decreases as a deposit is found and more details of the orebody become known.

Borrowing subsumes some asset that stands as surety for the loan. Since no known asset (orebody) exists when exploration begins, borrowed funds are clearly not an appropriate financial vehicle. Rather, exploration funding typically comes from monies that, if otherwise unspent, would be paid out as taxes. The financial risks remain high after an orebody is discovered since its dimensions must be delineated by drilling, tonnage and grade established, and the land position assured. Financing at this stage may be expected from a public stock offering, private venture capital, or transfer of the property to a major company while retaining a working interest or royalty; this latter being a common route for exploration companies. Monies raised at this stage provide for such needs as economic studies, planning of the mine, mill, and other installations, and acquisitions of essential permits.

Should the proposition remain viable to this point—a chance of but one in many hundreds—ordinary debt financing can be reasonably obtained based on the proven asset of the orebody and developed plans for its exploitation. Loans of this kind can be arranged with commercial banks, finance companies, insurance companies, development authorities, and governmental agencies.

For most programs, the capital requirements are those to initiate the program and decrease rapidly in time as the mine development is completed and the physical plant constructed. Operational costs must also be borne by capital outlays during this early period. Once operations begin, however, the cash flow from product sales should cover the operating expenses, gradually repay the borrowed capital and its debt service, pay dividends, and provide monies to a fund intended to capitalize the next venture.

In summary, the level of risk and the completeness of information available are the parameters controlling mine financing—risk falling as information rises. A brief checklist of essential information should include:

Company organization
Land status
Geology
Ore reserves
Proposed mining method
Proposed milling method
Permits from regulatory agencies
Markets

The Cost and Time Value of Money

Money, as any commodity, may be purchased, traded, or rented. The rental fee for money is called interest and is paid on a prearranged schedule at the rates established by lender policy or agreement between lender and borrower. The rate of interest will be affected by the creditworthiness of the borrower and the length of time for which the loan is to be in force. The components of a particular rate of interest may be identified as a core rate to which are added premiums for long-term lending risks, inflation, volatility, and fear.

The core or normal rate of return has historically been 2 to 3% and represents the willingness of investors and lenders to earn a profit equal to the average productivity gain in the overall economy. Premiums, usually 1 to 2%, are added to the core rate for longer term (more than 10 years) and thus riskier loans, and 1 to 3% might be added based on creditworthiness. A premium for inflation, up to 7% in 1982, is added to offset the anticipated decline in purchasing power over the life of the loan. Interest rates above the level of about 11 to 15% may be ascribed to premiums added to counter the volatility in the supply of money available for lending, about 2%, and simple fears of lenders regarding economic policy, another 2%.

Economic Considerations

Interest may be calculated daily, monthly, annually, or on any agreed time basis. The process is one of compounding whereby the interest earned in a previous period is added to the principal, thus increasing the base value on which the next calculation is made. For example, the growth of $100.00 at 10% annual interest is shown in Figure 8-3. These relations may be expressed by the single payment compound amount formula

$$S = P(1 + i)^n$$

where S is the sum of money at the end of n interest periods, P the original investment, and i the interest rate.

The present value of money or other assets receivable at some future time is called the discount rate and is the reciprocal of interest accumulation.

$$P = S/(1 + i)^n$$

The present value, P, is the payment required today for an asset receivable, S, after n interest periods at the discount rate i. Single payment present value factors for various values of n and annual compounding of interest are given in Figure 8-4. This information is usually presented in tables for greater accuracy, but a good approximation may be made from the diagram.

An important application of the use of discounting is the determination of the net present value or acquisition value of a property. Assuming estimates of costs, revenues, and reserves are reasonably accurate and that the purchasing power of the

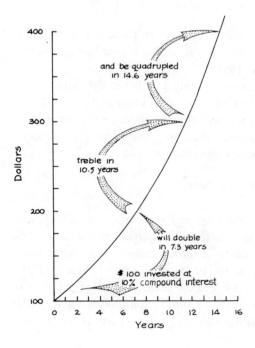

Figure 8.3. Growth of $100 at 10% interest.

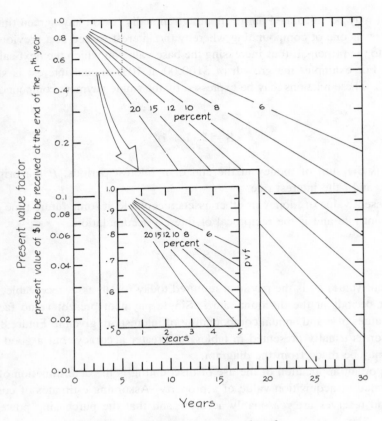

Figure 8.4. Discount rate or present value of money.

dollar will not significantly change, a balance sheet such as Table 8-1 may be set up based on a capital investment, say of $10 million, and a hurdle rate (threshold below which investment is unattractive) of 15%.

Mines and mills are designed to operate continuously at a particular capacity, so after the initiation of production or in the taking over of an ongoing operation, the income generated may be considered as a uniform flow of money over the mine's lifetime. A preliminary economic analysis may then take the form of an annuity whose present value is expressed by the series discount formula

$$P = A(1 + i)^{n-1}/i(1 + i)^n$$

where A is the annual income, i the interest rate, and n the number of payment periods. Figure 8-5 provides the present value factors for various interest rates compounded annually. As an example of the utility of these relations, consider a mine to be expected to earn $2.5 million per year for 10 years at a hurdle rate of 12%. The present value factor from Figure 8-5 is 5.6 and the present value of the income is thus 5.6 × $2,500,000 = $14,000,000. If the required capital investment is $10 million, then the net present value (acquisition value) of the property is $4 million.

Economic Considerations

Table 8-1.
Balance Sheet for Present and Acquisition Value

Year	Income	15% Discount Factor (from Fig. 8-4)	Present Value
1	$5 million	0.87	$4,350,000
2	5	0.76	3,800,000
3	3	0.66	1,980,000
4	2	0.57	1,140,000
5	2	0.50	1,000,000
6	3	0.43	1,290,000
Totals	$20 million		$13,560,000
		Less capital investment	10,000,000
		Acquisition value	3,560,000

Implications of a Wasting Asset

The principal financial distinction between a company engaging in the primary production of mineral commodities and other kinds of businesses lies in the fact that each unit of ore extracted reduces the material assets of the mining company. A business based on such a wasting asset must necessarily manage its affairs in a somewhat different manner from one dealing with renewable resources such as farming, timbering, or fishing or one engaged in manufacturing, transportation, or service.

Basically, the mineral-based company must accumulate sufficient monies over the lifetime of an operating property to fund the discovery and development or purchase

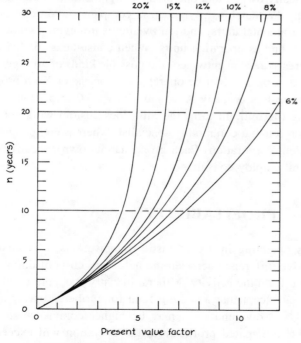

present value of annuity of $1 to be received at the end of each year for a period of n years

Figure 8.5. Annuity.

of a property whose value is at least equivalent to the initial worth of that which has been exhausted. For single mine companies this may be a one-for-one proposition, whereas for larger companies with many mines it may involve various trade-offs, always, however, leading to equivalency in worth. This need for asset replacement results in the continuing exploration, evaluation, option, and purchase activities carried on by mining companies.

The monies needed to acquire replacement assets are usually obtained by a modest assessment on the sales of each unit produced. This money is then invested at interest and is thus equivalent to a personal savings account set up for a college education or against a rainy day. Alternately, sums needed for acquisition may be borrowed.

OPERATING ORGANIZATIONS

The real and often high risks associated with the discovery, development, and operation of a mineral business often results in the sharing of risk—and anticipated profits—by a consortium. Such joint ventures become more common as the scale of the proposed operation increases and may lead to the collaboration of a number of corporate entities or private-government partnerships. Since these organizations may otherwise be in competition, it is usual practice to establish a new company for the project whose ownership is shared by the members of the consortium. Naturally, agreed arrangements for operational control and other responsibilties must be made.

Mineral deposits are typically owned and operated by organizations ranging from individuals to governments. Partnerships, syndicates, companies, corporations or corporate subsidiaries, governmental units, and joint ventures involving two or more organizations are all to be found as operating units. When consultants, partial ownership, and investment interests are also entwined, the possible kinds of arrangements are staggering. For example, a mine might be operated by contractor X on property of Y leased in a joint venture to Company A, a subsidiary of Corporation Z, and B, a syndicate also owning the mill. In turn, the mill does custom work for Corporation C, which markets through a commodity specialist, whereas company A has sales contracts with its parent corporation. Each group has its own executive hierarchy, investors, and several employ consultants.

PROFITABILITY

The heart of any business operating in a capitalistic framework is, of course, its profitability—the net positive difference between the necessary costs of doing business and the income derived from the activity. Mineral industry accounting will parallel that of other businesses except that a sinking fund for further property acquisition must be established and dividend rates may be higher commensurate with increased risks. Only if the anticipated profits from the operation will exceed the total costs by an amount adequate to offset all ordinary expenses plus those special needs for replacement should activity be initiated.

Economic Considerations

Ore and profit are inextricably linked and a mine may as well be "found" by an engineer or accountant who arranges matters to yield a profit as by a geologist who discovers the mineralization.* The same deposit might well be an ore in one instance and merely rock in another depending upon the way it is physically worked by the engineer or financially managed by the businessperson.

The various means for comparing investment proposals in mining are (1) the payback period, (2) net present value, and (3) discounted cash flow in investment. The payback period is simply the number of years needed for earnings to replace the original capital outlay and serves only as a coarse screen for making a judgment in that the time value of money is not considered.

The net present value of a property, discussed earlier, is the amount of money receivable in excess of the minimum acceptable rate of return on investment and is widely used as a component of decision making in the mineral industry. The interest rate chosen is based on the company's cost of capital, investments, and loans and usually includes factors related to the market value of the firm, its past record, and anticipated risk.

Discounted cash flow return on investment is basically concerned with identifying the rate of interest that will be received on the capital outlay. First the interest rate is found that balances the present value of income from the mine and the capital investment at a net present value of zero. For the case of uniform earnings, the payback period is found (investment divided by annual earnings), the lifetime assumed, and the return on investment interpolated from Figure 8-3. As an example, assume annual earnings of $750,000 on an investment of $5,000,000 and a lifetime if 15 years. The payback period (present value factor) is 5,000,000/750,000 = 6.67 years. From Figure 8-5, this present value factor gives an interest rate of 12.5% for a lifetime of 15 years.

PREDICTION AND RISK ASSESSMENT

Essential to any successful business is the ability to look into the future and anticipate those changes in supply and demand, market prices, cost of money, and the myriad variables that affect both short- and long-term financial health.

Mineral exploration and exploitation have for many an aura of romance stemming from tales of bonanza discoveries and huge profits. Unfortunately, although windfalls are certainly possible, they are highly unlikely and the mine operator or investor is well advised to undertake participation in the mineral industry under the same terms as any other business venture. If the romance adds a bit of sauce, so be it, but one should not use it in the balance sheet. The economics of a mining venture has all of the aspects of any business, being distinguished only by its foundation on a wasting asset that sets a finite life to the extraction and processing of the product.

A critical feature in any business is that all parameters are variable in time. Hence

---

*Too many "mines" have yielded a profit to operators of barren holes in the ground, although not to investors. Caution with respect to both chain letters and mining promotions is always advisable.

assessment of profitability in a mining business requires that predictions of both operating costs and market prices be made for the lifetime of the deposit. Prediction, which implies future knowledge of the date and amount or degree of some time-dependent parameter is, to say the least, a difficult task, but one that must be faced.

The general procedure in any prediction is to review the past performance of the particular parameter, usually by a plot of amount versus time, and extrapolate the trend of the curve into the future making such adjustments as experience suggests. The projection of a trend into the future may, however, not be easy as is suggested by Figure 8-6. The plot shows a general increase in amount (price, cost, tonnes, etc.) over time, but it is not clear whether at an increasing rate (curve a), a steady rate (curve b), or if the upward trend will level out (curve c). Experience and understanding of the dynamics of the particular situation must play a significant role is such projections.

Figure 8-7 is a real world example of a straight line projection for the market price of iron ore using data accumulated over a period of 70 years. Price paid per ton has been adjusted to the U.S. Wholesale Price Index for 1957–1959 = 100 and the trend line found by the least squares method. This effectively places the line in such a position that half of the values are above and half below it. A notable feature of the raw data for this curve is its large variability over short time-spans, a common phenomenon for some metal prices. This awkward feature may be reduced and the curve smoothed by the use of a running average. On the other hand, the details of variability provide some insight as to the risk to be assumed in accepting a particular prediction where risk is defined as the probability of change of a particular amount during a particular interval. Reference lines showing the percentage change from the price trend line have been added to Figure 8-7 and an analysis of the variance is given in Table 8-2.

Risk assessment is commonly employed by engineers to evaluate the incidence of natural disasters or other conditions that will affect their works—the "hundred-year flood" being a shorthand statement that flooding to a particular level has a recurrence interval of .01 years, that is, may be anticipated to occur once per 100 years. Note

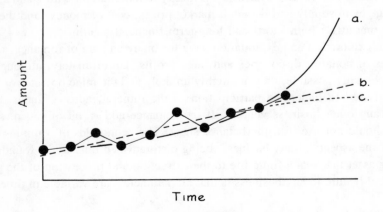

Figure 8.6. Extrapolation of a trend.

Economic Considerations

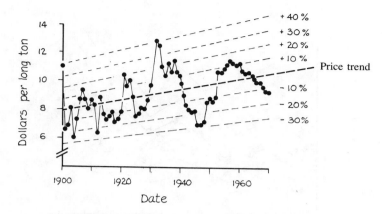

Figure 8.7. Iron ore price trend 1900–1970 (*Source:* Based on data reprinted with permission from J. S. Carman, *Obstacles to Mineral Development, a Pragmatic View.* Copyright © 1979, London: Pergamon Books Ltd.)

that the date at which the flood will occur is not addressed. Risk assessment may be used to evaluate the surety of a prediction of trends in costs and revenues, thus allowing some estimate of their probable correctness and assignment of a risk factor to a particular prediction.

The projection of aperiodic data such as commodity prices, wages, or interest rates into the future must be done in economic planning, but obviously entails risk. One approach to understanding and adjusting for this unclear future is described below and may be followed by reference to Figure 8-8. The first step is to plot the raw data for the past performance of the item of concern, in this instance an annual value over a 50-year span. Next, this irregular curve (dotted) is smoothed by an appropriate averaging method; in the figure this is done by plotting 5-year averages (solid line); the time for which the prediction is to be made. The smoothed curve may then be reasonably extrapolated using least squares analysis as shown by its dashed extension, which represents the anticipated average value projected into the future. Annual values in the past, however, have obviously deviated widely from the average and an assessment of this variation—the riskiness of the extrapolation—should be made.

The assessment of risk is done by adding curves of percent difference using the

Table 8-2.
Variance in Mesabi Non-Bessemer Iron Ore Prices, 1900–1970

% Change	Number of Times +	Recurrence Interval[1]	Number of Times −	Recurrence Interval[1]
40	1	0.014	0	—
30–40	3	0.042	0	—
20–30	7	0.100	5	0.071
10–20	17	0.243	28	0.400
0–10	34	0.487	37	0.528

[1]Number of times per year. A recurrence interval of 0.10 means once in ten years, 0.01 once in one hundred years, etc.

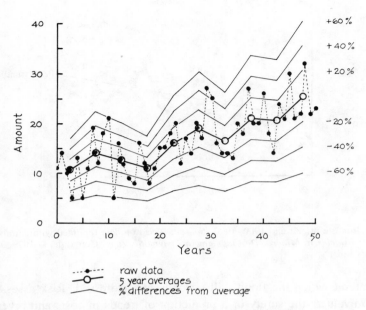

Figure 8.8. Risk assessment (see text for explanation).

smoothed curve as a reference line, which then shows the past performance of the raw data with respect to the smoothed. The number of the points in the zones defined by the lines of percent difference divided by the total time, 50 years, provides a recurrence interval for anticipated variation from the average. In the example, a change of up to 20%, either plus or minus, may be expected with a recurrence interval of 27/50 = 0.54, about every other year. Variation from the average between −20 and −40% may be expected 7/50 of the time, a recurrence interval of 0.14 or once in every 7.15 years.

The numerous economic parameters to be projected into the future will have interdependent values in an ideal economy in that inflation raises product prices along with operating costs and the cost of money is inversely related to its supply. The economy of mining and the world, however, is far from ideal because of such features as fixed price contracts, cartelization, wage and price regulations and agreements, protective tariffs, and war. In consequence, although some projected fluctuations may be offsetting, it is unlikely that all will be so mining must be considered a high-risk economic activity.

Projections of revenues and costs are often made with considerably greater accuracy than is warranted by the quality of the data available or certainty of future events; five-place discount tables provide impressive numbers but give no guarantee of reality if applied to an incorrectly estimated value. Table 8-3 shows the considerable effects of various modest errors in estimation.

The erratic behavior of metal prices suggested by the foregoing is certainly a real phenomenon and of serious concern for economic planning. Over long periods of time, however, prices have tended to have a relatively smooth rise closely related

Table 8-3.
Effects of Errors in Estimation

Revenue	100	95	90	85	80	
Cost	70	70	70	70	70	
Profit	30	25	20	15	10	revenue overestimated
Revenue	100	100	100	100	100	
Cost	70	75	80	85	90	cost underestimated
Profit	30	25	20	15	10	
Revenue	100	95	90	85	80	revenue over-
Cost	70	75	80	85	90	estimated, cost
Profit	30	20	10	0	−10	underestimated

The first column is an estimate of anticipated revenue, cost, and profit; other columns show the effect on profit of errors of estimation in revenue and cost.

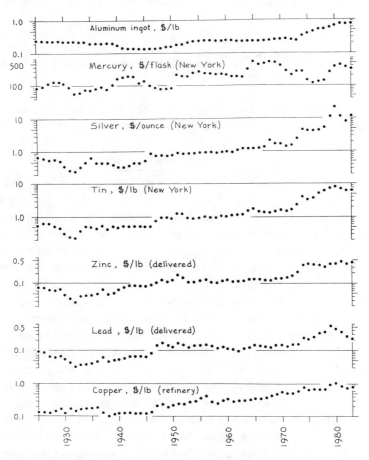

Figure 8.9. Average annual metal prices, 1925–1983 (*Source:* Data from *Engineering and Mining Journal.* Copyright © March 1984)

to the Consumer Price Index as indicated by Figure 8-9. Note that it is the logarithm of market price that has been plotted against time.

DETERMINATION OF LIFETIME

The lifetime of a particular mining activity has previously been singled out together with the tonnage and grade of an ore deposit as a critical determinant in establishing the technical forecast of mining rate and hence plant size. It is also apparent that financial planning must be similarly tied to anticipated lifetime.

The operating life of a property might be established in either of two ways. An executive decision made in the light of external parameters to exhaust the deposit in n years might be made, thus fixing the annual production rate and attendant income and costs

$$\frac{\text{Total tonnes in deposit}}{\text{Total years}} = \text{Tonnes produced per year}$$

However, it is more likely that an optimized mix of mining costs and mining rate will be sought and the lifetime of the operation determined by

$$\text{Lifetime} = \frac{\text{Total tonnes in deposit}}{\text{Tonnes mined per year}}$$

This is a particularly complex business problem since income and expenditure will vary in time and will probably never attain a steady state relationship. Figure 8-10 suggests the nature of the economic history of a mine and identifies the kinds of income and expenditures that must be balanced. The principal unknowns are the capitalization required, which will vary with the production scale according as the size of the surface plant, hoisting capacity, mining machinery, and other capital outlays, and the rate of income return from production as a function of lifetime. Debt service, operating expenses, profit margin, taxes, and related costs are variables dependent upon the amount of capitalization and production rate. All, of course, are modified by changes in time of the cost of money, wages, and prices.

A solution may be approximated by assuming a particular capitalization level that identifies the amount of debt service and fixes the mining rate (with its attendant direct costs) and hence the lifetime. Forecasts of income increments because of price changes (+ or −) may then be applied to the revenue derived from product sales.

From mining rate and lifetime, predictions of operating costs (wages, maintenance, supplies), taxes, replacement costs, and profits, all adjusted for time, are then possible and a lifetime balance sheet can be set up for a particular trial case. The exercise is then repeated several times including both larger and smaller rates than

Economic Considerations

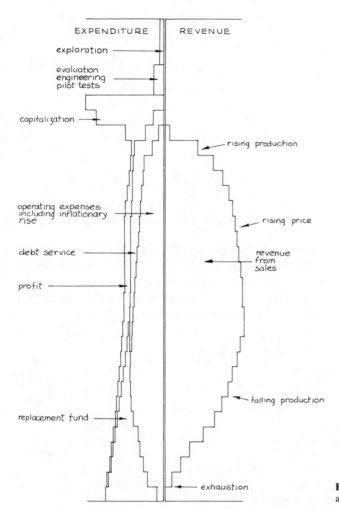

Figure 8.10. Cash flow model for a mining venture.

experience suggests as optimum to generate an estimation of profit margin for each trial case and the results plotted as in Figure 8-11. Obviously, the mix with the greatest profit margin will be chosen. In modern planning of mine operations computer programs are used to randomize cost and revenue figures for each phase of the operation and for several projected production rates, which, when compared, allows the most profitable scale of operations to be found.

A NUMERICAL EXAMPLE

Financial planning for a mining venture is based on features of the orebody and cost analyses generated by the geologist and engineer. Since 10's to 100's of millions of dollars are typically involved, it may be anticipated that this will be a lengthy and

Figure 8.11. Optimization of profitability.

exhaustive exercise whose complexity increases with the size of the project. Fortunately, as may be seen from the following example, the data, which is a mix of known and estimated factors, lends itself readily to computerization using an appropriate algorithm. Programs surveying the effects of changes in the various parameters used with a goal of optimization should be a part of any feasibility study.

Table 8-4.
Capital Cost Summary

Mine	Mine equipment	$ 594,450	
	Underground development	2,694,000	
			3,288,450
Mill	Mill equipment	1,222,100	
	Installation	120,000	
	Building construction	700,800	
			2,042,900
Plant and Services	Powerhouse equipment, construction, and installation	795,000	
	Water supply and sewage equipment and installation	170,200	
	Tailings disposal	102,000	
	Heating plant for mine	118,800	
	Office, changehouse, and service building	328,500	
	Vehicles and site roads	80,000	
			1,594,500
	Grand total	6,925,850	
	Contingencies @ 5%	346,000	
	Engineering @ 2% of mine and 10% of plant costs	270,000	
	Interest at 9% on preproduction capital of $5,000,000	450,000	
	Inventory	190,000	
	Working capital and operating cost for 4 weeks	210,000	
	Overall capital cost	$7,841,850	

Table 8-5.
Operating Cost Summary

Mine	Mine personnel and fringe benefits	5.77 $/tn.s	
	Explosives	0.34	
	Supplies	2.16	
			8.27 $/tn.s
Mill	Mill personnel and fringe benefits	3.21	
	Supplies	1.43	
			4.64
Plant	Personnel	3.57	
	Supplies	0.84	
	Operating costs of power and heating plants	1.72	
			6.13
	Overall operating cost		19.04 $/tn.s

The following example has been inspired by a real situation and has been deliberately selected as a small, short-lived, single mining venture. This helps to reduce the complexity of the financial analysis and is, presumably, a more likely activity for the interested nonspecialist.

The Marginal copper mining property in Canada has been explored by surface geological mapping, some underground development, and diamond drilling from both the surface and underground. Some $1,700,000 has been spent on this exploration

Table 8-6.
Revenue Summary

1. Proven ore reserve = 736,000 short tons less 10% for mine support, losses, and uneconomically located ore	662,400 tn.s
2. Mine operation, 260 days/year, 2 shifts/day, 740 tn.s/day	192,500 tn.s/yr
3. Mill operation, 350 days/year, 3 shifts/day	192,500 tn.s/yr
4. Mill feed grade	3.2%
5. Concentrate grade	26%
6. Beneficiation ratio; line 5/line 4	8.125
7. Recovery of copper from concentrate	93%
8. Tonnes of concentrate produced each day; (550 from line 3/line 6) × line 7	62.95 tn.s
9. Copper content of concentrate per day; line 8 × 2000 × line 5	32,734 lbs
10. Smelting losses @ 20 lb/ton of concentrate; 20 × line 8	1,259 lbs
11. Net daily copper production; line 9 − line 10	31,475 lbs
12. Market price of copper; assumed	$0.80/lb
13. Smelter settlement @ 95% of market price; line 12 × line 11 × 0.95	$23,921
14. Less smelter charges @ $27/tn.s;	$1,700
15. Less trucking of concentrate @ $20/tn.s; $20 × line 9	$1,259
16. Net daily smelter revenue; line 13 − (line 14 + line 15)	$20,962
17. Net revenue per ton produced; line 16/550 tn.s from line 3	$38.11/tn.s
18. Direct operating costs for mining, milling, and plant operation; from Table 9–5	$19.04/tn.s
19. Operating profit; line 17–line 18	$19.07/tn.s
20. Average annual operating profit; line 3 × line 19	$3,670,975

Table 8-7.
Financial Analysis

1. Lifetime (mineable ore/mining rate; 662,400/192,500)	3.44 years
2. Annual operating profit (from Table 8-6)	$3,670,975
3. Total capital cost (from Table 8-4)	$7,841,850
4. Plus prepaid development expenditure	$1,700,000
5. Less inventory and working capital (from Table 8-4)	$300,000
6. Depreciable capital expenditure; line 3 + line 4 − line 5	$9,241,850
7. Annual depreciation over 3.44 years; line 6/line 1	$2,686,584
8. Taxable income; line 2 − line 7	$984,391
9. Taxes at 50% of taxable income; .50 × line 8	$492,196
10. Annual cash flow; line 2 − line 9	$3,178,779
11. Total cash flow over mine lifetime; line 10 × line 1	$10,935,000
12. Present value of $1.00 annually at 12% for 3.44 years; Figure 8-5	2.682
13. Present value of cash flow for 3.44 years; line 12 × line 2	$9,845,555
14. Annuity computation: Ratio of project costs to annual operating profit inclusive of prepaid development expenditures; line 3/line 2	2.136
Corresponding annuity of $1.00 discounted over 3.44 years	25.1%

Table 8-8.
Cash Flow Summary (000's omitted)

	Year					
	1 (6 mos.)	2	3	4	5	6
Tons of ore produced			192.5	192.5	192.5	84.9
Net smelter return, $			7336	7336	7336	3236
Operating cost, $			3665	3665	3665	1616
1. Operating profit			3671	3671	3671	1620
2. Development cost			—	—	1700[1]	—
3. Interest			448	318	188	58
4. Pretax profit			3223	3353	1733	1562
5. Provincial tax			170	170	170	72
6. Profit before federal tax			3053	3183	1613	1490
7. Depreciation			—	—	1613	1490
8. Depletion			—	—	—	—
9. Taxable income			—	—	0	0
10. Income tax			—[2]	—[2]	0	0
11. Balance (9 − 10)			—	—	0	0
12. Net income (2 + 7 + 8 + 11)			3053	3183	3313	1490
13. Income to dividends			1256	1415	1570	654
Debt drawdown	2714	2286				
Equity drawdown		5400				
Debt outstanding			5000	3547	2094	641
Debt repayment (1/3.44 per yr)			1453	1453	1453	641
Debt remaining at year end			3347	2094	641	0

[1] Deferred exploration and development.
[2] Tax free period.

program, which has developed 736,000 tonnes of ore having an average grade of 3.2% copper. Engineering studies have developed an optimum mining and milling plan that will mine 740 tonnes of ore per day using two shifts and mining for 260 days per year and will mill 550 tonnes each day with three shifts operating for 350 days per year. The annual production of the mine and mill is thus 192,000 tonnes per year. The mill concentrate will be shipped to a custom smelter for recovery of copper metal from the sulfide ore concentrate.

Tables 8-4 to 8-8 are summaries of anticipated capital cost, operating cost, revenue, financial analysis, and cash flow for the project.

References

Bierman, H., and Shmidt, S. 1966. *The Capital Budgeting Decision,* 2d ed., New York: Macmillan.

Lewis, F. M., and Bhappu, R. S. 1975. Evaluating mining ventures via feasibility studies. *Mining Engineering* 27, no. 10, 50–55.

O'Neil, T. J. 1974. The minerals depletion allowance: Its effect on future supply and financing. *Mining Engineering* 26, no. 11, 39–41.

Peters, W. C. 1987. *Exploration and Mining Geology,* 2nd ed. New York: John Wiley & Sons.

Van Horn, J. C. 1968. *Financial Management and Policy.* Englewood Cliffs, NJ: Prentice-Hall.

Chapter 9
The Legal Framework

BACKGROUND FOR MINERAL LAW

Mineral extraction and processing must be conducted in a manner consistent with the applicable statutes and guidelines promulgated at every level of government. Although these may appear, or even be, unnecessarily restrictive, they have been developed with the common good in view. Unfortunately, local and state regulations differ so much in detail that no short review is possible, and even at the national level the number of applicable rules is unmanageable except in broad outline.

Because mineral exploration involves the direct use of materials of the Earth's surface or subsurface, it is essential that ownership be clearly established. Although it is mining that is being considered, the nature of mineral ownership is basically that of any property. Title resides in some individual, private group, or governmental body and, consistent with the pertinent statutes of the body politic, rights may be sold, rented, or conveyed by gift or otherwise. Thus owners may separate and sell rights to mine all or specified minerals, enter into option agreements, arrange for rental (lease, concession, royalty), or otherwise exploit the value of their property.

Owners, other than a dictatorial central government, do not, however, have unlimited freedom in the disposal of their rights in land. Land use, environmental, or other statutes will generally bar uses detrimental to neighbors or making significant deleterious environmental impact. Further, special statutes control a landowner's rights to water on or beneath a property. In mosts eastern states water may be used on the property but must not be modified in quality or quantity to a downstream neighbor. Riparian laws rule in the West whereby usage allotments have been established by priority or agreement.

Two fundamental principles form the basis of all U.S. mining laws, namely (1) the right of mine operators to secure an indefeasible title to their property so long as they fulfill certain specified conditions, compliance with which is absolutely within their power, and (2) the right of a state or other landlord to certain rents, royalties, or taxes on the profits of mines and to a reasonable performance of work.

The mining laws of the world fall into two broad classes: (1) the concession system in which a private owner or, more usually the state, holds title and has the right to grant concessions or leases, and (2) the claim system wherein any individual has the

The Legal Framework 183

right to hold, work, or dispose of certain limited areas of ground for its contained minerals, subject to certain specified restrictions.

The concession system arose as a consequence of feudalism, which vested rights to property in feudal lords and kings and, with modifications, the system is used today in countries or their former colonies with a feudal history. In practice today these rights have become vested in central governments and concessionary arrangements are under the control of some designated agency. A significant feature of the concession system is that the state or other owners select concessionaires and thus may pursue an integrated mining policy. Advantages, from the point of view of the state, are the placement of the properties in the hands of dependable operators with consequent maintenance of order and regular payment of royalties (The origin of this term is thus obvious) coupled with the ability to exercise some control over the capitalization and financial transactions of the concession holder. The principal difficulties are the possibility of favoritism allowing the concentration of unduly large powers over mining property in a few individuals, the tendency to hold unworked ground, and the reduction of free and innovative competition.

The claim system originated in the United States during the California Gold Rush of 1849 when discoveries were made on the just-acquired public lands. Arrangements to identify a particular area of ground that a claimant was allowed to work and the conditions for holding it had to be made on the spot so as to maintain public order. Thus began the system of staking and recording claims that has been developed and codified for public lands in the U.S. as well as South Africa and Canada.

LANDS IN THE PUBLIC DOMAIN

A feature of the founding of the United States was that property rights in the territory of the 13 original states (and Vermont, Kentucky, Maine, and West Virginia, which were later carved out of them) remained with the states. The federal government acquired no property rights in this territory. Rapid acquisition of federal lands was initiated between 1781 and 1802 when the area of the present states of Tennessee, Illinois, Indiana, Ohio, Michigan, Wisconsin, and portions of Alabama and Mississippi was ceded to the federal government by the original states. The Louisiana Purchase of 1803 added 2.6 million square kilometers to the public domain and extended the territorial limits of the United States westward to the Rocky Mountains. Texas was annexed in 1845 and retained ownership of its public domain. Expansion to the Pacific was accomplished in 1848 by the treaty with Mexico of Guadalupe Hidalgo, and the Gadsden Purchase of 1853 fixed the present border between the United States and Mexico. Both of these actions with Mexico involved the obligation on the part of the United States to protect valid titles acquired under Mexican rule. In 1898 by treaty of cession between the United States and Hawaii, all land in the Hawaiian Islands except private lands passed to the United States. Puerto Rico was ceded to the United States by Spain in 1898 and all crown lands passed to the United States. Soon after, in 1902, these lands were ceded by the United States to the Territory of

Puerto Rico. Public lands in Alaska were acquired by the United States with its purchase from Russia in 1867 and have been retained by the federal government. The public lands of the United States are shown in Figure 9-1. Their concentration in the western third of the country should be noted.

The current use of land in the United States, exclusive of offshore acreage, is shown in Figure 9-2. About 750 million acres (one third of the total) is public land, mostly under the management of the Bureau of Land Management, U.S. Forest Service, and the National Park Service. There has been an accelerating trend in recent years to withdraw public lands from access. Beginning with the Wilderness Act of 1964, about 70–75% of public lands have been closed or restricted for mineral exploration and development. The impact of these actions has been particularly severe in Alaska as shown by Figure 9-3.

U.S. MINING LAW

Some basic tenets of the mining law of the United States are that mining is ranked as a private industry to be fostered and regulated as any other industry, a grant or conveyance by the federal government carries all minerals unless expressly reserved, no royalties are reserved, and upon conveyance mining land becomes private property subject to the same rules of law as other real property.

There is no provision made in federal law for the location of claims on privately owned lands, and mineral development thereon is solely a matter for private negotiation. Transfer of property rights among private owners follows the accepted prac-

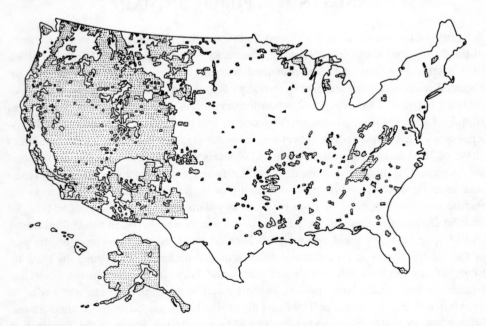

Figure 9.1. Public lands in the United States (*Source:* From *U.S. News and World Report.* Copyright © February 13, 1984)

The Legal Framework

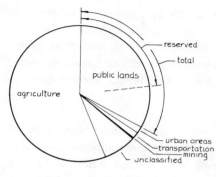

LAND DISTRIBUTION IN THE UNITED STATES

Figure 9.2. Land distribution in the United States (*Source:* Data from American Institute of Professional Geologists, *Metals, Minerals, Mining.* 2/e rev. Copyright © Golden, 1981)

tices of private rights—purchase arrangements by mutual agreement including the separation and reservation of surface rights, mineral rights, and right of access. Claims may be staked on public lands of some states, whereas concessionary arrangements are in force for the state-owned lands of others.

There are three major mining laws that authorize and state the policy of the federal government with respect to private exploration and development of minerals on public lands in the United States:

1. The General Mining Law of 1872, "An act to promote the development of the mining resources of the United States," declares that valuable minerals on public lands are open to private exploration and purchase. This statute, with a few additions and amendments, provides the framework for the acquisition of title to mineral property by private parties through patent by staking, recording, and performing assessment work on claimed areas. The size of individual claims is established as being parallelograms with sides of respectively 600 and 1500 feet.

2. The Mineral Leasing Act of 1920 was enacted in recognition of the impracticality of working small claims for bulk mineral commodities and authorized the federal government to act as a lessor to private parties for deposits of coal, oil, gas, oil shale, phosphate, and sodium (salt) on public lands. Later amendments added potash, sand, clay, gravel, stone, and sulfur.

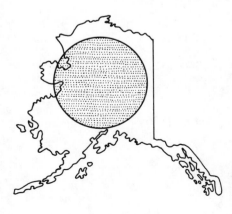

Figure 9.3. Approximate amount of Alaskan land area withdrawn from mineral development (*Source:* Data from American Institute of Professional Geologists, *Metals, Minerals, Mining,* 2/e rev. Copyright © Golden, 1981)

3. The Mining and Minerals Policy Act of 1970 (Public Law 91-631) states that "The Congress declares that it is the continuing policy of the Federal Government in the national interest to foster and encourage private enterprise in the development of economically sound and stable domestic mining, minerals, metal, and mineral reclamation industries. . . ." Unfortunately, this clear policy has not been vigorously pursued.

Set against these three enabling laws are numerous statutes that restrict mineral exploration and development on certain public lands and environmental laws that limit and control mining irrespective of its location in the United States. The major federal actions that limit mineral development on public lands are:

Forest Service Organic Act of 1897
Forest and Rangeland Renewable Resources Planning Act of 1974
National Forest Management Act of 1976
The Federal Land Policy and Management Act
Reclamation Act of 1902
Wilderness Act and related Executive Orders
The Wild and Scenic Rivers Act
National Trails System Act
Coastal Zone Management Act
Protection of Wetlands (Executive Order)
National Wildlife Refuge System Administrative Act of 1966
Endangered Species Act of 1973
Fish and Wildlife Coordination Act
The Antiquities Act
Archeological and Historical Preservation Act
National Historic Preservation Act of 1966
Alaska Native Claim Settlement Act of 1971

The environmental laws, an uncoordinated group of special purpose statutes to protect and improve the quality of the air, water, land, and aesthetic values of the overall environment are:

National Environmental Policy Act of 1969
Clean Air Act and Clean Air Act Amendments of 1977
Federal Water Pollution Control Act
Water Quality Improvement Act of 1970
Clean Water Act Amendments of 1977
Safe Drinking Water Act
Federal Resource Conservation and Recovery Act of 1976
Noise control Act of 1972
Toxic Substances Control Act
Uranium Mill Tailings Radiation Control Act of 1978

Surface Mining Control and Reclamation Act of 1977
Refuse Act of 1980

Such restrictive statutes are not unique to the United States; legislation confining mining to approved methods and in approved places is to be encountered worldwide. Obviously, competent legal advice regarding ownership, mining statutes, and environmental or other inhibitory legislation should always be sought for any contemplated mining activity.

TAXATION

The revenue needed to support the wide-ranging functions and services of government is largely supplied by taxes levied on its citizens and those companies doing business within its jurisdiction. Taxes are an inevitable part of the cost of operating any business, and the mineral industry is no exception. The three bases of taxation commonly used for levies in the mineral industry are income, production, and property.

Income taxes, as for individuals or nonmining businesses, are based on net profit and in the United States are levied by both federal and state governments. This is, however, the only direct federal tax on mining. Further, special provisions in the U.S. Internal Revenue Code regarding the extraction of natural resources are designed to deliver a subsidy to mining activities. This is done by allowing exploration and development costs to be deducted rather than capitalized and recognizing an orebody's wasting nature by according an allowance for depletion.

The depletion allowance, unique to the mineral industry, is an annual deduction from gross income provided to mineral producers with the intent of providing them with monies to discover another orebody to replace the ore being exhausted. The deduction may be either by cost (unit depletion) or by the more usual percentage depletion. Cost depletion allows the total cost of acquisition of a mineral property to be divided among the tonnes of ore as they are mined. Percentage depletion permits deduction of a specified percentage of the gross income providing the amount is less than 50% of the before-depletion net income.

Royalty payments are excluded from the gross income of a company in the determination of the depletion allowance. The right to claim depletion is passed on to the royalty holders who are expected to claim their share. Depletion allowances in force in 1986 for domestic production and for foreign production by domestic companies are given in Table 9-1.

Taxes on production are levied by individual states and may be termed severance taxes or production royalties. They are levied regardless of the profit or loss sustained by the company in the taxable period and are unique to mineral production. Such taxes are not, for example, imposed on the production of automobiles, ketchup, or pulpwood.

Property taxes, sometimes termed ad valorem taxes, are levied by state and local

Table 9-1.
Depletion Allowances, 1986

Commodity	D[1]	F[2]	Commodity	D	F	Commodity	D	F
Aluminum	0%	0%	Gold	15%	14%	Rhenium	14%	14%
Antimony	22	14	Graphite	14–22	14	Rubidium	14	14
Arsenic	14	14	Gypsum	14	14	Rutile	22	14
Asbestos	22	10	Hafnium	22	14	Salt	10	10
Barite	14	14	Ilmenite	22	14	Sand	5–14	5–14
Bauxite	22	14	Indium	14	14	Scandium	14	14
Beryllium	22	14	Iodine	14	14	Selenium	14	14
Bismuth	22	14	Iron ore	12	11.2	Silicon	5–14	5–14
Boron	14	14	Kyanite	22	14	Silver	15	14
Bromine	5	5	Lead	22	14	Soda Ash	14	14
Brucite	10	10	Lime	14	14	Sodium sulfate	14	14
Cadmium	22	14	Lithium	22	14	Stone, crushed	5–14	5–14
Cesium	14	14	Magnesium chloride	5	5	Stone, dimension	14	14
Chromium	22	14	Manganese	22	14	Strontium	22	14
Clays	5–22	5–14	Mercury	22	14	Sulfur	22	22
Cobalt	22	14	Mica	22	14	Talc	14–22	14
Columbium	22	14	Molybdenum	22	14	Tantalum	22	14
Copper	15	14	Nickel	22	14	Tellurium	14	14
Corundum	22	14	Olivine	22	14	Thallium	14	14
Diamond	14	14	Peat	5	—	Thorium	14–22	14
Diatomite	14	14	Perlite	10	10	Tin	22	14
Dolomite	14	14	Phosphate rock	14	14	Titania	—	—
Feldspar	14	14	Platinum group	22	14	Tungsten	22	14
Fluorspar	22	14	Potash	14	14	Vanadium	22	14
Gallium	—	—	Pumice	5	5	Vermiculite	14	14
Garnet	14	14	Quartz crystal	22	14	Yttrium	14–22	14
Gemstones	14	14	Rare earth metals	14–22	14	Zinc	22	14
						Zirconium	22	14

[1]Domestic.
[2]Foreign.
Source: Data from U.S. Bureau of Mines.

governments. The determination of the tax to be paid, as for any property tax, requires that a market value, tax rate, and assessment rate be established. Since the market value of a mine requires information about the ore in place as well as the surface plant, its determination will be quite artificial. Because the tax includes the value of proven but unmined ore, a common effect is for the mining company to limit exploration and development work and to keep ore in sight for only a few years ahead. This has the interesting side effect that the amount of proven reserves of a commodity may suggest that we will run out of it within a decade.

Figure 9-4 illustrates the impact of these different kinds of taxes on mineral production and the amount of tax which each generates. It is assumed in the figure that the production rate and production costs are constant and that only the market price varies in time. A severance tax is applied as a flat rate per unit of production, an ad valorem tax as a shrinking amount with time as the orebody is depleted (10% per year in the example), and an income tax as a percent of profit, 50% in the example. The profit before tax, the same in each instance, and the amount of tax paid (stippled) is tabulated for each time period and after-tax profit calculated. On the assumptions used, only an income tax would allow the company to remain in business. Other tax rates would, of course, modify the results, but obviously severance and ad valorem taxes make their greatest impact on profitability at times of depressed markets, whereas taxes on income rise with profitability.

Studies of the effects of the kinds of taxes levied and the taxation rates used in base metal mines in Canada (MacKenzie and Bilodeau, 1979) have shown that taxation on profits (income tax) rather than gross revenue (a function of production and

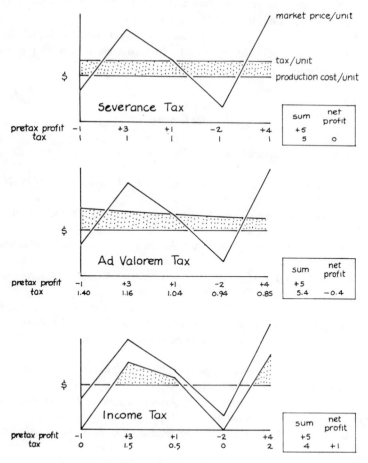

Figure 9.4. Impact of different kinds of taxes.

Table 9-2.
Effects of Taxation on Mining in British Columbia

	With Royalties	Without Royalties
Present value to society	$29 million	$116 million
Rate of return to investors	6.6%	8.8%
Number of economic deposits	12	15
Present value of tax payments	$592 million	$710 million

Source: American Inst. of Professional Geologists, *Metals, Minerals, Mining* 2/e rev. Golden, 1981.

thus a severance tax) is 2 to 10 times better in present value to society, improves the rate of return to investors one and a half to two times, and more than doubles the number of deposits that can be expected to be profitable. The effect of the repeal of production royalties on mines by the provincial government of British Columbia, Canada, in 1976 is a case in point. The striking results are shown in Table 9-2.

Other instances of increased mine production following changes to a more favorable tax climate may be seen in recent developments in Ireland, Argentina, and Jamaica. Per contra, mineral-rich "Bolivia is a prime example where a lack of flexibility and unwillingness to address the adverse effects of an antiquated tax law have severely hurt the mining industry which, in turn, has had negative effects on the economy as a whole. . . ." (DAP, 1980).

References

AIPG Colorado Section Special Issue. 1981. *Metals, Minerals, Mining*, 2d. ed., Golden, CO: AIPS.

Cameron, E. N., ed. 1973. *The Mineral Position of the United States, 1975–2000.* Madison: University of Wisconsin Press.

DAP, 1980. Taxes on mining can be dangerous and backfire on the governments concerned. *World Mining*, 33, no. 5, 98.

Leshey, J. D. 1987. The Mining Law: A Study in Perpetual Motion. Washington DC: Resources for the Future.

Lincoln Institute of Land Policy. 1977. *Non-renewable Resource Taxation in the Western States*. Lexington, MA: D. C. Heath.

Mackenzie, B. W., and Bilodeau, M. L. 1979. *Effects of Taxation on Base Metal Mining in Canada*. Kingston, Ont: Centre for Resource Studies, Queen's University.

Maley, T. S. 1985. *Mining Law: From Location to Patent*. Boise, ID: Mineral Land Publications.

O'Neil, T. J. 1974. The minerals depletion allowance: Its effect on future supply and financing. *Mining Engineering*, no. 10, 61–64.

Outerbridge, C., ed. (Updated irregularly). *American Law of Mining*. Bender, NY: Rocky Mountain Mineral Law Foundation.

Peele, R., and Church, J. A. 1941. *Mining Engineers Handbook*, 3d. ed. New York: John Wiley & Sons.

Chapter 10

Mineral Resources and Trade

RESOURCES AND RESERVES

The mineral deposits of the world are not all discovered and those that are known are often incompletely evaluated. However, geologic insight allows estimates of the world totals for various commodities to be made and, even though these estimates may differ by factors of 3 to 10, they indicate the dimensions of the ultimate potential *resource* base of a particular commodity. This total includes presently subeconomic and undiscovered mineral deposits together with discovered and evaluated orebodies.

The U.S. Geological Survey employs a classification illustrated by Figure 10-1 to assess mineral resources on the basis of the level of knowledge of their existence and magnitude together with the feasibility of their economic and technological recovery. Deposits may be identified or undiscovered and may be recoverable, paramarginal (recoverable with a modest improvement in economic level), or submarginal (conceivably recoverable). Deposits that have escaped discovery in known mineralized districts are termed hypothetical resources, and those that could occur in untested ground where conditions are favorable are termed speculative.

Only a small portion of the mineral deposits of the Earth are currently identified, measured, and evaluated to such an extent that they can be designated as ore, that is, worked at a profit. The level of confidence of knowledge of quantity and quality for identified and recoverable resources—*reserves*—is indicated by the terms proven, probable, and possible. The classification of ore into these categories is not fixed but will be continuously changing in a particular deposit as mining proceeds with concomitant exploration, development, and re-evaluation.

Historically, resource and reserve estimates have been overly conservative. In 1947 the U.S. Geological Survey and U.S. Bureau of Mines, based on known deposits, their rate of consumption, and methods of recovery, estimated that the following commodities would now be exhausted:

aluminum	copper	lead	sulfur
bismuth	fluorite	petroleum	vanadium
cadmium	gold	silver	zinc

191

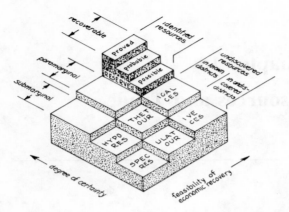

Figure 10.1. Categorization of resources and reserves. (*Source:* Based on U.S. Geological Survey classification)

Obviously, this has not happened! Figure 10-2 shows the changes in published reserves for the 20-year period from the late 1940s to the late 1960s. The impression gained from these records is that resources will be adequate for needs well into the twenty-first century, although it remains true that demand is increasing exponentially and the absolute resource base is finite.

The mineral position of the United States has undergone a marked transformation since World War II. U.S. ore production prior to that time accounted for about one-half of the world production for 18 important commodities (Fig. 10-3) and generally provided for U.S. needs. World production has risen steeply since then and far outstripped that of the United States. Further, U.S. consumption has become greater than production requiring increasing importation of raw materials.

Resource availability for the United States has been categorized by the U.S. Geological Survey by estimating the *minimum anticipated cumulative demand*—MACD—for the period 1975–2000 (Cameron, 1973). Resource availability compared with the MACD is grouped into six classes in Table 10-1, and the values for a number

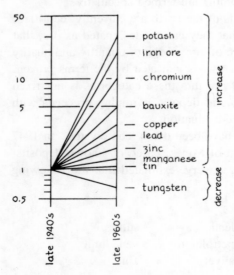

Figure 10.2. Changes in published reserve estimates. (*Source:* Based on data from reprinted with permission from J. S. Carman, *Obstacles to Mineral Development: A Pragmatic View*. Copyright © 1979, London: Pergamon Books Ltd.)

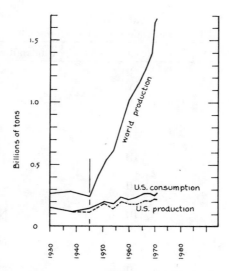

Figure 10.3. World production and U.S. production and consumption of eighteen mineral commodities (iron ore, bauxite, copper, lead, zinc, tungsten, chromium, nickel molybdenum, manganese, tin, vanadium, fluorspar, phosphate, cement, gypsum, potash, sulfur). (*Source:* From E. N. Cameron, ed. *The Mineral Position of the United States. 1975–2000,* Copyright © 1973, Madison: University of Wisconsin Press)

of commodities are listed in Table 10-2. Note that a commodity classed as III or greater must be imported to meet current needs.

THE INTERNATIONAL VIEWPOINT

The erratic distribution of the Earth's mineral wealth often means that the economics of a particular deposit must be considered using an international viewpoint; thus concern with means and cost of overseas transport, foreign policy and politics, customs duties, currency exchange rates, and related matters are essential components of the mining picture. The distribution of the world of mineral deposits, market, and available risk capital is such that seldom are all three found within the same national framework. The so-called developed countries are the principal sources of investment capital through private and corporate sources or the financial support of international

Table 10-1.
Resource Availability

Class	Times MACD
I	10 or more
II	2–10
III	0.75–2
IV	0.35–0.75
V	0.10–0.35
VI	less than 0.1

Source: U.S. Geological Survey

Table 10-2.
MACD Classification of Some Commodities

Commodity	Reserves	Unidentified Resources
Aluminum	II	NE[1]
Asbestos	V	VI
Barite	II	II
Chromium	VI	V
Copper	III	III
Fluorine	V	V
Gold	III	NE
Gypsum	I	I
Iron	II	I
Molybdenum	I	I
Nickel	III	NE
Phosphate	II	I
Sand and Gravel	III	NE
Sulfur	I	I
Titanium	II	II
Uranium	II	III
Vanadium	II	NE
Zinc	II	II

[1]Not estimated for various technical reasons.
Source: U.S. Geological Survey

lending organizations. However, the principal mineral resources of these countries, although extensive, are inadequate in kind and amount to supply all of their requirements and they must turn to less-developed countries for additional supplies.

Less-developed countries often see their mineral resources as national treasures that can be used as a steppingstone for entry into a modern technologic world. Capital and technical knowledge must, however, be imported in order to exploit the income-producing raw materials. Thus there is generated a mutual need for international interaction with a concomitant flow of money, goods, and services. Unfortunately, the fact of mutually recognized needs does not automatically generate a mutually beneficial trade.

Consider first the point of view of mining companies in the United States, Japan, Canada, or Great Britain. Their primary function is not to mine but to make a profit, which they do by supplying certain mineral materials to consumers in their home country or elsewhere in the world. To do this they must have mineral deposits in production and be able to realize a satisfactory return on their investment. Their ore may be in one foreign country and principal market in another, but most of their investors are at home so a profit must be returned to the home country. Obviously, working agreements regarding the conduct of business, tariffs, the export of capital, foreign equity and employees, and so on must be made abroad and the present and future political climate of the country in which investment is contemplated becomes a significant factor in decisions regarding foreign operations.

World events, perhaps local in nature but worldwide in impact, are too common to overlook in planning and operations. Within the past few years civil strife in Zaire has disrupted copper and cobalt production with consequent skyrocketing of cobalt prices; mining in the western Saharan phosphate area was shut down due to annexation of the area by Morocco; iron and diamond mining in Liberia and tin mining in Bolivia were curtailed by revolution; and diplomatic ties between the United States and China opened trade in tungsten graphite and antimony.

The point of view of the country in which a mineral deposit occurs must be understood with sympathy. The deposit may be the country's only significant asset and certainly it wants to obtain every possible advantage of income, training of personnel, and development of infrastructure through agreements with extranational companies. Typical of such agreements are provisions for the construction of roads, schools, and hospitals, technical training of nationals on site or by study abroad, limitation of the number or percentage of foreign employees, majority national ownership, marketing restrictions, and limitations on the export of profits.

In a few instances, the emphasis on national interest by resource-rich countries has brought a number together with the intent of controlling the supply and price of a particular commodity. Best known is OPEC (Organization of Petroleum Exporting Countries), a consortium of 13 oil-producing countries. A similar organization of tin-producing countries, the International Tin Committee, included almost all of the major tin-producing nations until its collapse in 1987. The Association of Iron-Ore Exporting Countries was organized in 1970 with Peru, Mauretania, Algeria, Venezuela, Chile, India, and Australia as members. Other groups include the Intergovernmental Council of Copper Exporting Countries, CIPEC, (Chile, Peru, Zambia, Zaire), the International Bauxite Association (Australia, Guinea, Guyana, Jamaica, Sierra Leone, Surinam, Yugoslavia), and 11 tungsten-producing countries. The supply and price of diamonds in world trade has been successfully controlled by the deBeers organization for many years; producers outside the group have, so far, been satisfied with the prices as established.

Over time the resolution of the different views of supplier and consumer nations has brought about a complex interplay in the worldwide flow of money and materials including the establishment of favored-nation partnerships, banking relationships, and technical exchange.

The world population is doubling every 35 years and estimates of mineral consumption indicate a doubling rate of 25 years or less; hence the exploration for primary mineral resources must be continuously and exponentially expanding even in the face of increasing efforts in conservation, recycling, and substitution. The result, a steady depletion of these wasting assets worldwide, regional and national exhaustion of supply, increases in price, and shifting of sources. However, much of the difficulty experienced by suppliers is not a lack of raw material but their inability under restrictive regulation or excessive taxation to discover and produce it. The following excerpt from *Metals, Minerals, Mining* (1981) expresses the concern of one professional group of the "mineral world," the American Institute of Professional Geologists, and indicates (Fig. 10-4) the dependence of the United States on world trade for a long list of mineral commodities.

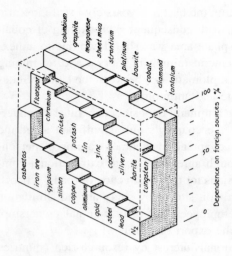

Figure 10.4. Net import reliance of the United States on some non-fuel mineral commodities as a percentage of apparent consumption, 1986. Other mineral commodities having a significant degree of import reliance are antimony, bismuth, gallium, iodine, ilmenite, mercury, rhenium, rutile, and vanadium. (*Source:* Data from U.S. Bureau of Mines)

Once the most powerful nation in the world, the United States has lost the respect and much of the leadership it formerly enjoyed. Many factors have contributed to this, but major ingredients are the dependence of the country on foreign energy and mineral sources and the worsening metals shortage. After decades of heedless energy consumption, the U.S. now finds itself critically short of petroleum, and worse, indecisive as to how best utilize its vast resources of coal and uranium.

Where do we stand with regard to self-sufficiency in metals, and why is mining so important to our economy? During World War II and as recently as 1979, various federal statutes authorized stockpiling of certain metals deemed as strategic, necessary for defense, and in short domestic supply. As many as 79 metals and minerals have been stockpiled to provide from 1 to 5 years' supplies, for defense purposes only. In recent years the availability of some of these metals—chrome, manganese, cobalt, and platinum group metals—to name a few—is becoming less secure because of our total reliance on imports. Figure 10-4 shows this dependence on some selected minerals and metals. Availability of many metals can be constricted by relations with such nations as Turkey, Rhodesia, South Africa, and Bolivia, major suppliers of chrome, cobalt, tin, manganese, and tungsten. Our stockpile supplies of cobalt, platinum group metals, chromium, and titanium are dangerously low.

Recent confrontations have prompted a new look at our military capabilities. Stockpiles cannot presently furnish the necessary critical metals. The impending shortage of these materials which are vital components of household appliances, automobiles and even bicycles, will also affect our everyday lives.

Coupled with the immediate need for strategic materials is the long term necessity for a sound and healthy domestic mining industry. Minerals and agricultural products have long been the foundations of a solid economy and national security. Mining in other countries is encouraged and even subsidized, but the mining industry in the U.S. is subject to increasing taxation, duplicative networks of state and federal regulations, and the preservationist movement.

The third facet of the metals problem is the distribution of metalliferous deposits and legislation and administrative regulations governing exploration, development, and mining on public lands. Many types of mineral deposits are located in belts or regions of

Mineral Resources and Trade

geologic disturbance—where the earth's crust has been broken and buckled, in some areas by intrusions of molten rock, in other areas by deep-seated compressive stresses. Those same geologic forces that govern the locations of mineral deposits, also provide rugged topography, high mountains and multi-colored rock patterns that produce beautiful scenery. . . .

The crisis in metals is a compound problem that has been growing in many directions over a period of years. It has been nurtured by neglect and by short sightedness. It has been exacerbated by those in our society who would impose a philosophy of no growth under the guise of environmentalism by forcing legislation to curtail existing mining operations and prevent new plant construction, and by encouraging federal withdrawal of lands from responsible mineral exploration.

Figure 10-5 provides a synoptic view of the distribution of commodity sources for the United States in 1986. Canada, Mexico, Brazil, Republic of South Africa, and China are major suppliers of a number of the needed mineral commodities and, when the sources of a particular commodity is examined, the view of a limited number of suppliers developed in Chapter 8 quickly emerges. For example, those commodities on which the United States is completely dependent are:

Commodity	Major Sources, 1982–1985
Columbium	Brazil, Canada, Thailand, Nigeria
Graphite	Brazil, Mexico, China, Malagasy
Manganese	Brazil, Republic of South Africa, Gabon, France
Sheet mica	Japan, Belgium, India, France
Strontium	Mexico, Spain

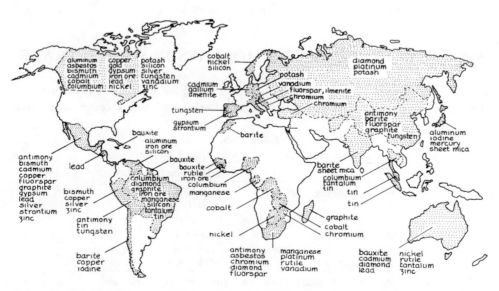

Figure 10.5. Major suppliers of critical mineral raw materials to the United States, 1986. (*Source:* Data from U.S. Bureau of Mines)

TRANSPORTATION

Each commodity may be expected to have its own peculiarities of source, quality, marketing, capitalization, and regulation, but one constant is that trade requires transport, and transport cost is a function of means and distance. Some comparative costs are given in Table 10-3 and, although these are 1972 data and higher today, the relative differences are not significantly changed.

To a first approximation, shipping costs may be taken as increasing linearly with distance (Fig. 10-6), and the millhead value of a commodity will be reduced in a reciprocal manner. Materials of low unit value, curve 1, cannot be economically shipped very far from their source so a given deposit has its local market insulated by distance from competitors. The commodity is said to have a high place value. Examples are sand and gravel, brick clays, and crushed rock.

Commodities of high unit value, often traded in smaller tonnages, may be economically transported over longer distances, curve 2. When the distance limit for such materials is equal or greater than the longest shipment distance on the Earth, it becomes freely tradable.

Shipping costs must also include the handling costs associated with any change of carrier and may be taken to include tariffs or duties, although these should be otherwise accounted for. The transport cost curve for Figure 10-6 might thus be modified as shown in Figure 10-7 where the steps represent cost increments due to transshipment with no increase in transport distance.

Obviously, every effort should be made to reduce direct shipping costs by use of the most economical combination of carriers and, when possible, unit trains, special cargo ships, and special means such as overhead tramways or slurry lines. Similarly, automatic loading and unloading facilities should be utilized.

SOME EXAMPLES OF COMMODITIES IN WORLD TRADE

The variety of mineral commodities in world trade and the special circumstances that apply to each preclude any exhaustive description; rather, a few commodities that

Table 10-3.
Comparative Costs of Commodity Transport

Means	Cents per Tonne per Kilometer
Ocean shipping	0.02–0.7
Pipeline	0.10–0.7
River barge	0.15–0.3
Railroad	0.28–1.1
Truck	3.9–5.0
Air freight	8.4–14.0

After P. J. Maddex. How "changing" transportation can affect mining profits. (New York: *Am. Inst. Mining Metall. Petroleum Engineers.* SME Preprint 72-H-78, 1972).

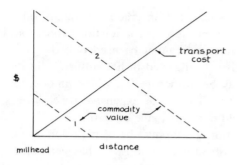

Figure 10.6. Commodity value and transport cost.

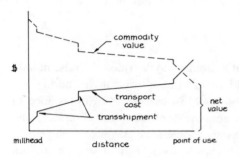

Figure 10.7. Effect of transshipment on commodity value and transport cost.

sample the principal commodity categories and exhibit different market characteristics are highlighted here.

To set the stage, it is useful to realize that mineral commodities represent 30–35% of all commodities traded and that the annual value of all crude mineral commodities in terms of constant dollars has increased four times in the past 30 years to over $200 billion. When the value added by processing and transportation from producing areas is included, the annual value of mineral commodities is of the order of $500,000 million.

Phosphate Rock

Phosphorus is a crucial ingredient of fertilizer and its principal source is phosphate rock; 143 million tonnes were mined worldwide in 1981 with 73% produced in the U.S.S.R., the United States, and the Saharan region, especially Morocco. Over half of the production traded in the free world was from the southeastern United States and Morocco, with Russian exports essentially limited to eastern Europe.

The global reserves of phosphate rock are variously estimated at 100 to 1,000 billion tonnes excluding other significant sources of phosphorus in apatite concentrations in certain igneous rocks and guano. Reserves are completely adequate. Extensive deposits in the United States are worked in Florida and the largely undeveloped Phosphoria Formation covers 259,000 km^2 in six northwestern states. Aside from its primary phosphorus content, phosphate rock contains small amounts of uranium and fluorine, which may be recovered as byproducts.

There are more known phosphate rock deposits in the world than are currently in production, a number in regions where the need for crop fertilization is great. However, the majority of these deposits are either small, low in grade, or contain substantial amounts of materials deleterious in presently used beneficiation processes. Phosphate rock as mined is relatively insoluble and cannot be used directly as a fertilizer. To make it suitable for use, it is usually digested in hot sulfuric acid to make superphosphate, $H_4Ca(PO_4)_2$, which may be further treated with ammonia to yield ammonium superphosphate, $H_4NH_4(PO_4)_2$. This is an expensive process requiring about 1 tonne of sulfur as acid to yield one-half tonne of fertilizer.

The present requirement that both ore and finished product be of high quality in order to hold processing and transportation costs at economic levels has generated the trade pattern in which exports are limited to hard currency purchases.

Bauxite

Bauxite, the ore of aluminum, is a mixture of aluminum hydroxide minerals, mainly gibbsite, $Al(OH)_3$, which forms under conditions of intense tropical weathering and is found today in areas of present or past tropics. Most bauxite (89%) is refined to aluminum metal and the remainder used in abrasives, refractories, ceramics, and various chemical reagents. Refining is by the energy-intensive Bayer process; leaching under high pressure and temperature with caustic soda to yield soluble sodium aluminate from which aluminum oxide (alumina) is precipitated followed by electrolysis at about 1000° C with molten cryolite, $NaAlF_4$, to produce aluminum metal. Five to 7 tonnes of bauxite, 23 kg of fluorine, and 7 to 8 KWH of electrical energy are needed to make 1 tonne of metal. Because of the high energy requirements, ore is typically shipped to localities with cheap hydroelectric power sources such as Canada or Norway, or to those having wellhead natural gas for refining.

Bauxite has a low unit value so mining operations are restricted to large deposits of high purity in order to offset the high shipping and refining costs associated with aluminum production. Fortunately, many such bauxite deposits are known and many are not in production, for example, extensive orebodies in Guinea and the Amazon region of Brazil. Other possible ores of aluminum include high-aluminum clays, nepheline syenite, and anorthositic rocks. If oil shale is developed in the Colorado-Wyoming area, there will probably be some byproduct aluminum production since the rock contains dawsonite, $NaAl(CO_3)(OH)_2$, which is readily soluble and transformed to sodium aluminate.

Nearly 90 million tonnes of bauxite were produced by 27 countries in 1980 with Jamaica (14%), Guinea (15%), and Australia (31%) as the leaders. Just under 33 million tonnes of alumina were produced in 25 countries with the United States and Australia accounting for 40%.

The U.S. consumption of bauxite in 1980 was 15.6 million tonnes, one-third of the world production, of which only 10% was domestically produced and the remainder imported from the Caribbean area (Jamaica, Guyana), West Africa (Guinea), and Australia. The international aspects of the aluminum business are underlined by a new smelter in South Carolina owned jointly by American Metals Climax (AMAX), Mitsui and Co., Ltd., and the Nippon Steel Company, whose ore is provided by the

Aluminum Company of America (Alcoa) from Jamaica and mines of Australian Proprietary, Ltd., in northern Australia.

Tungsten

An extremely high melting point, 3370° C, coupled with high hot strength and hardness, make tungsten metal and its alloys with carbon and iron essential for drilling bits, machine tools, machinery, and other applications where resistance to wear and abrasion or operation at high temperature is important.

The principal ore minerals of tungsten are wolframite, $Fe(WO_4)$, and scheelite, $Ca(WO_4)$, both rare in the United States. The most important deposits are in geologically young mountain belts and 60% of the world reserves of 2 million tonnes of metal are located in a zone extending from northern Korea through southeastern China to northern Malaysia.

The trading units for tungsten ore and concentrate are either short tons of 60% tungsten trioxide (WO_3) or pounds of contained metal. The total free world production in 1979 was just over 100 million pounds (about 50,000 tonnes) of which the United States mined only about 1.5%, mainly from four western mines. Principal imports of ore and concentrate to the United States were from Canada, (3.1 million pounds), Bolivia (3.0 million pounds), Thailand (1.2 million pounds), and mainland China (1.2 million pounds). Tungsten is considered a strategic metal by the U.S. government because of its direct military uses and is indirectly of military importance because of its essential role in machine tools and oil drilling. Strategic materials are stockpiled by the federal government against a military crisis and to provide a buffer against artificial short-term variations in the market. At the beginning of 1980 this stockpile had goals and on-hand stocks, not all of which were of adequate grade, as shown in Table 10-4.

The critical nature of tungsten to U.S. technology is suggested by the modification by Executive Order in 1979 of the Generalized System of Preferences so that tungsten ore, concentrate, ferrotungsten, ferrosilicon tungsten, and waste and scrap from designated beneficiary countries could be imported duty free. These materials nominally have import duties from Most Favored Countries of $0.17/lb tungsten content for ore and concentrate, $0.21/lb + 6% ad valorem import duty for ferroalloys, and $0.73/lb for waste and scrap.

Table 10-4.
Tungsten Stockpile, 1980
(thousands of pounds)

	Goals	On Hand
Concentrate	8,823	91,817
Ferrotungsten	17,769	2,025
Metal	3,290	1,899
Tungsten carbide	12,845	2,033

Source: U.S. Bureau of Mines

Platinum Group Metals (platinum, palladium, iridium, osmium, rhodium, ruthenium)

Platinum, whose value exceeds that of gold, is certainly a precious metal, but whereas nearly half of the gold produced is made into jewelry, only 3% of platinum is so used. Most platinum and palladium, the principal members of the platinum metals group, are consumed in industrial applications dominated by catalytic uses in the automotive and petroleum industries. The use of platinum and palladium in the United States is indicated by Figure 10-8.

In 1980 about 6.8 million troy ounces of platinum group metals (21.1 tonnes) were produced worldwide, 48% in the U.S.S.R., 45% in the Republic of South Africa, and 6% in Canada. The United States imported over half of this production (about 3.5 million troy ounces) and contributed less than 1% of its needs from domestic mines.

The very limited production of platinum group metals (and nickel) in the United States reflects the paucity of nickeliferous ultramafic rocks, which are their principal source. Placers derived from peridotite intrusions at Goodnews Bay, Alaska, and some byproduct recovery from copper ores was the only U.S. production until 1987 when mining began in the Beartooth Mountains of Montana. The ore is in the Stillwater Complex, a layered ultramafic igneous intrusion about 45 kilometers long and 4–6 wide. Three mineralized zones are recognized; the Basal Series containing copper-nickel sulfides, the Ultramafic Series containing concentrations of chromite, and the platiniferous Banded Series, 1 to 3 meters thick and 35 kilometers long. The ore in this zone has a grade of about 0.8 ounces per ton and a platinum-palladium ratio of $1:3$.

Some $45 million has been invested to date in the venture by a consortium of Chevron Resources Company (managing partner), the Manville Corporation, and Lac Minerals, Ltd. of Canada. Mining is by cut and fill stoping and concentration by froth flotation. Because there is no adequate smelter in North America, the concentrate is being shipped to Belgium for refining. The planned mining rate is 1,000 tpd yielding 200,000 ounces of combined metals per year. The lifetime is estimated at 20–30 years.

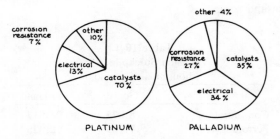

Figure 10.8. Uses of platinum and palladium in the United States, 1980. (*Source:* From U.S. Bureau of Mines, *Minerals Yearbook*. Copyright © 1980)

Zirconia

Zirconium, a minor metal, is a component of the minerals zircon, $ZrSiO_4$, and baddeleyite, ZrO_2, both of which are heavy and resistant. Although sparsely distributed in primary igneous and metamorphic rock, these minerals together with a number of others, including important rutile and ilmenite (titanium-bearing) and such rare earth element minerals as monazite and xenotime, have been concentrated in beach placers at a number of localities. These mineral sands—"black sands"—are widely distributed, but most of the economic deposits are in the southern hemisphere where they are worked in Australia, mainland China, India, Malaysia, Sri Lanka, Thailand, and the Republic of South Africa. Northern hemisphere production is from the United States and U.S.S.R. Titaniferous minerals are usually the principal product derived from such black sands with zirconium minerals recovered as a coproduct.

Zirconia (zirconium oxide) and zirconium metal are highly refractory materials and nearly half of the zircon and zirconia produced is used as foundry sand for high temperature casting. The remainder is used in a multitude of applications including coatings for welding rods, refractory brick, ceramics, abrasives, paints, opacifier for enamelware, polishes, and toughener for rubber. Zirconium metal is mostly used as fuel rod containers in nuclear reactors with minor uses in refractory alloys, corrosion-resistant equipment, and photo flash bulbs.

Australia is the major world producer of zircon from mineral sands on both the east and west coasts. Reserves are estimated at 14 million tonnes of which 20% is unavailable because of environmental considerations. An additional 5 million tonnes are paramarginal. Zirconia exports from Australia in 1980 amounted to slightly more than 500,000 tonnes with most going to Japan (186,000 tonnes), the United States (98,000 tonnes), and Italy (87,000 tonnes). In addition to these imports, the United States produced a zircon coproduct from dredging and milling mineral sands in Florida.

Mica

Mica exhibits a unique combination of properties, notably its ability to cleave into thin, tough, strong, and flexible sheets having a high dielectric constant, low heat conductivity, high temperature resistance, and chemical inertness that makes it essential to the electrical industry. In finely ground form it is used in a variety of ways such as a filler, rubber dusting powder, lubricant, and adding luster to wallpaper, or it can be mixed with adhesives and molded.

Mica is a bewildering commodity in that the term in the marketplace includes only some of the minerals of the mica family (muscovite, phlogopite, and minor biotite and sericite plus some synthetic material) and excludes others such as vermiculite, lepidolite, and roscoelite. Further, the better quality mica shares with gemstones a complex system of grading, mostly by eye, in which size and perfection play important roles. Finally, mica is marketed in a wide range of physical forms intended for numerous end uses.

Broadly, there are four forms in which mica enters international trade; *block* representing unsplit books, *split* (splittings, thins, sheets, films), which is its more familiar form, *ground* either wet or dry, and *waste,* scrap, or flake. Built-up products combining ground or flake mica with various adhesives are in common use. The complexity of the grading of mica is indicated by the use by the Indian Standards Institution of 16 categories of visual quality classification for muscovite books ranging from "ruby clear" to "green/brown, stained" and 13 size categories ranging from 5 to 630 cm^2, each having a specified minimum dimension of one side of a usable rectangle (Murthy, 1964).

The production of block and sheet mica is highly labor intensive. Large books of mica are found only in pegmatite bodies, the more common muscovite in quartzose pegmatites with worldwide distribution and phlogopite in quartz-free pegmatites in Canada and Madagascar. These bodies are mined by small-scale methods and the ore is cobbed, cleaned, trimmed, split, and graded by hand. Flake muscovite or biotite may be recovered from mica-rich igneous or metamorphic rocks by more conventional mining and milling techniques.

World trade in unmanufactured, and cut, punched, ground, and built-up forms of mica is also complex. The principal producers of block, sheet, and scrap mica are India, U.S.S.R., Madagascar, and Korea. The United States dominates the production of ground mica, principally from mines in North Carolina but also in nine other states. The import-export pattern for the United States in 1985 is shown in Table 10-5.

Gold

Because of its intrinsic beauty and unique properties, and because demand has always exceeded supply, gold has been used since the stone age as a visible expression of wealth and basis of currency. Because of these uses, its recovery presents an unusual economic situation since the value of gold in the marketplace is not fixed by production costs and, as a result, the grade of workable ore varies in time.

Gold is widely distributed in both space and time and, in primary lodes, is generally found in hydrothermal deposits closely associated with igneous intrusions of salic or intermediate composition. A possible exception to this origin is the world's largest producing district, the Witwatersrand of South Africa, considered by many to be a fossil placer with some degree of hydrothermal reworking. A recent study, however, classifies the deposit as hydrothermal (see Chapter 3). Although gold is widely distributed in nature, its production, like that of most mineral commodities, is dominated by a relatively few countries. Figure 10-9 shows the principal producers in 1986.

Gold is enriched in the late hydrous fraction during the crystallization of magma and believed to be carried in hydrothermal solutions as an aurous chloride complex.

Mineral Resources and Trade

Table 10-5.
U.S. Imports and Exports of Mica, 1985 (values in thousands of pounds)

Type	Imports	Exports	Principal trading partners
Ground, pulverized		14,460	Canada, Mexico, Venezuela, Spain
	12,097		Canada
Waste, scrap		2,918	Mexico, Canada
	7,960		India
Unmanufactured block, film, splittings	1,684	82	India
Manufactured cut, stamped, built-up	909	5,103	Canada, Mexico Belgium, Netherlands, Japan
Miscellaneous	69		
Totals	22,719	22,563	

Source: Data from *Minerals Yearbook* vol. 1, U.S. Bureau of Mines, 1985.

Precipitation of gold from such solutions would occur on cooling, reaction with wall rocks, or by contact with carbonaceous materials. Because of its chemical inertness and malleability, native gold is essentially unchanged during its release by weathering and in transport. Its high specific gravity (19.3) causes it to travel in streams at much different rates than other rock and mineral particles and to be concentrated on bedrock in placer deposits. The origin of these deposits may be better understood if it is realized that in flood stage when the sediment in a stream channel is thrown into suspension, the heavy gold particles respond more slowly to suspension and more rapidly to settlement.

Native gold is not chemically pure, but always alloyed with silver, copper, or other elements to levels ranging from about 1 to 50%. Copper imparts a redness and gold becomes paler with increasing silver content. Extended solid solutions of gold with silver (electrum), copper (auricupride), palladium (porpezite), and rhodium (rhodite) are known but, except for electrum, are very rare. The purity of gold is

Figure 10.9. World gold production, 1986. (*Source:* Data from U.S. Bureau of Mines)

expressed as *fineness* in parts per thousand (pure gold = 1,000 fine) or in carats with pure gold = 24. Other units and conversions used in discussions of gold are:

Troy pound = 12 troy ounces.
Troy ounce = 1.09714 avoirdupois ounce = 31.104 grams.
Pennyweight (dwt) = 0.05 troy ounce = 1.5552 grams.
Part per million (ppm) = 1 gram per metric ton (tonne).

Exploration for gold deposits is carried on by the traditional method of panning, by the use of geochemical pathfinders such as arsenic or the arsenic-antimony ratio, by geophysical magnetic methods to locate associated magnetite in placers, and by remote sensing to discover fracture systems and mineralogic changes related to the passage of hydrothermal solutions. The techniques for the extraction of gold ores vary widely because of the variety of gold deposit types and include most of the methods described in Chapter 6, especially open pit and underground mining and dredging. Recovery of gold from its ores usually is done by cyaniding in tanks or heaps as discussed briefly in Chapter 7.

The reason for the current high interest in gold mining is, of course, the dramatic rise in the price of gold by a factor of more than 10 in the past 20 years. A tenfold rise in price greatly improves the profits of on-line mines and allows orebodies only one-tenth as rich to be mined; the cutoff grade for underground mining is now about 5 grams per tonne.

The position of the United States with respect to gold has been different from that of other nations. The United States was the only nation to retain the gold standard after 1919 and, until the late 1960s, was the only country in which anyone could dig for gold without government control. Prior to 1934 the world price of gold was secured at $20.67 per fine ounce by price guarantees of the U.S. government so dentists, jewelers, investors, and governments could purchase gold at that price (plus commission) from traders. In 1934, however, the money supply was expanded by depreciating the dollar value of gold to $35. At the same time, all bullion in private hands was transferred to the Treasury.

From 1938 to 1948 the U.S. gold reserves rose from $14.6 to $24.4 billion, two-thirds of the world's total monetary gold. Much of this huge reserve was used in the conduct of the Marshall Plan during the 1950s to revive the economies of European countries. By the 1960s the United States was spending more dollars abroad than could be justified by $35 gold and some countries, notably France and Switzerland, regularly cashed in dollars for bullion. The U.S. reserve fell to $12.1 billion.

The demand for gold in the 1960s grew to such an extent that the pressure on international currencies caused considerable maneuvering by the International Monetary Fund, representing countries with gold-owning central banks, in an attempt to maintain a controlled price for gold. In 1971 the United States, the principal holder of gold reserves, unilaterally stopped trading dollars for gold with governments and central banks. Gold prices began an immediate rise and appear to have stabilized in the $400 range, although fluctuating significantly in response to market pressures.

Mineral Resources and Trade

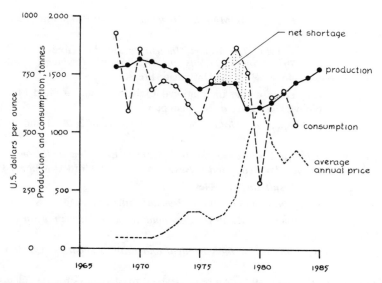

Figure 10.10. World production, consumption, and price of gold, 1968–1985. (*Source:* Data for the plot taken with permission of the publisher from *World Gold Deposits* by J. J. Bache, Copyright © 1987 by Elsevier Publishing Co., Inc.)

The roughly parallel changes in the price of oil over the same time-span may be noted. OPEC pricing raised the cost of oil by about the same factor as free-market trading raised gold values; the two are thus, to a large degree, offsetting.

The effects of these market changes are shown in Figure 10-10. Surprisingly, world production has actually fallen in the two decades since its price was freed even though at least 86 gold mines worldwide were either planned or actually placed in production in the years 1977–1987. The net shortage in supply with respect to demand from 1976 to 1980 resulted in the steep rise of gold prices to an historic high at $843 in mid-January 1980.

References

Bache, J. J. 1987. *World Gold Deposits; A Geological Classification.* New York: Elsevier.

Barney, G. O. 1984. *The Global 2000 Report to the President.* New York: Pergamon Press.

Blunden, J. 1985. *Mineral Resources and Their Management.* New York: Longmans.

Bosson, R., and Varon, B. 1977. *The Mining Industry and the Developing Countries.* New York: Oxford University Press.

Boyle, R. W. 1987. *Gold: History and Genesis of Deposits.* New York: Van Nostrand Reinhold.

British-North American Committee 1976. 1976. Mineral Development in the Eighties: Prospects and Problems. British-North American Committee

Cameron, E. N. 1973. *The Mineral Position of the United States, 1975–2000.* Madison: University of Wisconsin Press.

———. 1986. At the Crossroads. New York: John Wiley & Sons.

Carman, J. S. 1979. *Obstacles to Mineral Development, a Pragmatic View*. London: Pergamon Press.

Colorado School of Mines. 1958. *The Mineral Industries Bulletin*, Golden.

Fishman, L. L. 1980. *World Mineral Trends and U.S. Supply Problems*. Baltimore: Resources for the Future, Johns Hopkins University Press.

Foster, R. P., ed. Newsletters of the International Liaison Group on Gold Mineralization. Department of Geology, University of Southampton.

Green, T. 1987. *The Prospect for Gold: The View to the Year 2000*. London: Rosendale Press.

Harris, D. P. 1984. *Mineral Resources Appraisal: Mineral Endowment, Resources, and Potential Supply: Concepts, Methods, and Cases*. New York: Oxford University Press.

Industrial Minerals. *Metal Bulletin Inc.*, New York.

International Gold Mining Newsletter. *Mining Journal Ltd.*, Edenbridge, Kent.

Jensen, M. L., and Bateman, A. M. 1981. *Economic Mineral Deposits*, 3rd ed. rev. New York: John Wiley & Sons.

Johnstone, S. J. 1961. *Minerals for the Chemical and Allied Industries*. London: Chapman & Hall.

Journal of Economic Geology. 1975. An issue devoted to platinum group elements. vol. 71, no. 7.

Leontief, W., and others. 1983. *The Future of Non-fuel Minerals in the U.S. and World Economy. Input–Output Projections 1980–2030*. Lexington MA: Lexington Books.

McDivitt, J. F., and Manners, G. 1974. *Minerals and Men*. Baltimore: Johns Hopkins University Press.

McKelvey, V. E. 1973. Resources. Mineral Resource Estimates and Public Policy in United States Mineral Resources. Geological Survey Professional Paper 820.

Mertie, J. B., Jr. 1969. Economic Geology of the Platinum Metals. U.S. Geological Survey Professional Paper 630-A.

Metal Bulletin. *Metal Bulletin Inc.*, New York.

Mineral Industries Bulletin. Colorado School of Mines Research Institute, Colorado School of Mines, Golden.

Minerals Yearbook, 1985. Vol. I, Metals and Minerals. Washington, DC: U.S. Government Printing Office.

Murthy, M. V. N. 1964. *Mica Fields of India*. New Delhi: International Geological Congress.

Pearson, C. S., ed. 1987. *Multinational Corporations, Environment, and the Third World: Business Matters*. Durham, NC: Duke University Press.

Roush, G. A. 1939. *Strategic Mineral Supplies*. New York: McGraw-Hill.

Sheldon, R. P. 1982. Phosphate rock. *Scientific American*, vol. 246, no. 6.

Sutlov, A. 1971. *The Soviet Challenge in Base Metals*. Salt Lake City: The University of Utah Printing Services.

Warren, K. 1973. *Mineral Resources*. New York: Penguin Books.

Wise, E. M. 1964. *Gold, Recovery, Properties, and Applications*. New York: Van Nostrand Reinhold.

Wolfe, J. A. 1983. *Mineral Resource Perspectives*. London: Chapman & Hall.

———, 1984. *Mineral Resources: a World Review*. London: Chapman & Hall.

Yih, S. W. H. 1979. *Tungsten: Sources, Metallurgy, Properties, and Applications*. New York: Plenum Press.

Appendices

I Atomic parameters
II Mineral determinative tables
III Market specifications
IV Units and conversions

Appendix I

Atomic Parameters

Element	Symbol	Atomic Number	Atomic Weight,[1] $^{12}C = 12.000$	Usual Valence[2]	Usual Coordination[2]	First Ionization[2] Potential, eV
Actinium	Ac	89	227.03	3+		6.9
Aluminum	Al	13	26.98	3+	4,6	5.98
Antimony	Sb	51	121.75	3+,5+	6,4	8.64
Argon	Ar	18	39.95	0		15.76
Arsenic	As	33	74.92	3+,5+	6,4	9.81
Barium	Ba	56	137.33	2+	8,12	5.21
Beryllium	Be	4	9.01	4+	4	9.32
Bismuth	Bi	83	208.98	3+	6,8	7.29
Boron	B	5	10.81	3+	3,4	8.30
Bromine	Br	35	79.90	1−	6	11.84
Cadmium	Cd	48	112.41	2+	6,8	8.99
Calcium	Ca	20	40.08	2+	6,8	6.11
Carbon	C	6	12.01	4+	3	11.26
Cerium	Ce	58	140.12	3+	6,8	5.60
Cesium	Cs	55	132.91	1+	8,12	3.89
Chlorine	Cl	17	35.45	1−	6	13.01
Chromium	Cr	24	51.99	3+,6+	6,4	6.76
Cobalt	Co	27	58.93	2+	6	7.86
Copper	Cu	29	63.55	1+,2+	8,6	7.72
Dysprosium	Dy	66	162.50	3+	6,8	6.80
Erbium	Er	68	167.26	3+	6,8	6.08
Europium	Eu	63	151.97	2+,3+	6,8	5.67
Fluorine	F	9	18.99	1−	6	17.42
Gadolinium	Gd	64	157.25	3+	6,8	6.16
Gallium	Ga	31	69.72	3+	4,6	5.97
Germanium	Ge	32	72.61	4+	4,6	7.88
Gold	Au	79	196.97	1+,3+	8,12,4	9.22
Hafnium	Hf	72	178.49	4+	6,8	—
Helium	He	2	4.00	0	0	24.48
Holmium	Ho	67	164.93	3+	6	—
Hydrogen	H	1	1.008	1+	1,2	13.60
Indium	In	49	114.82	3+	6,8	5.79
Iodine	I	53	126.90	1−	6,8	10.45
Iridium	Ir	77	192.22	2+,4+	6	—
Iron	Fe	26	55.85	2+,3+	6,4	7.87
Krypton	Kr	36	83.80	0	0	14.00

Element	Symbol	Atomic Number	Atomic Weight[1], $^{12}C = 12.000$	Usual Valence[2]	Usual Coordination[2]	First Ionization[2] Potential, eV
Lanthanum	La	57	138.91	3+	6,8	5.61
Lead	Pb	82	207.21	2+	6–12	7.42
Lithium	Li	3	6.94	1+	4	5.39
Lutetium	Lu	71	174.97	3+	6,8	—
Magnesium	Mg	12	24.31	2+	6,8	7.64
Manganese	Mn	25	54.94	2+,3+,4+	6	7.43
Mercury	Hg	80	200.59	2+	6,8	10.43
Molybdenum	Mo	42	95.94	4+6,+	6,4	7.10
Neodymium	Nd	60	144.24	3+	6,8	5.51
Neon	Ne	10	20.18	0	0	21.56
Nickel	Ni	28	58.69	2+	6	7.63
Niobium	Nb	41	92.91	5+	4,6	6.88
Nitrogen	N	7	14.01	5+	3	14.53
Osmium	Os	76	190.20	4+	6	8.50
Oxygen	O	8	15.999	2–	4,6,8	13.61
Palladium	Pd	46	106.42	4+	4	8.33
Phosphorus	P	15	30.97	5+	4	10.48
Platinum	Pt	78	195.08	2+,4+	6,4	9.0
Polonium	Po	84	209	4+	8	8.43
Potassium	K	19	39.10	1+	8,12	4.34
Praesodymium	Pr	59	140.91	3+	6,8	5.46
Protactinium	Pa	91	231.04	4+	8	—
Radium	Ra	88	226.03	2+	8,12	5.28
Radon	Rn	86	222	2+	8,12	10.75
Rhenium	Re	75	186.21	4+	6	7.87
Rhodium	Rh	45	102.91	3+,4+	6	7.46
Rubidium	Rb	37	85.47	1+	8,12	4.18
Ruthenium	Ru	44	101.07	3+,4+	6	7.36
Samarium	Sm	62	150.36	3+	6	5.6
Scandium	Sc	21	44.96	3+	6,8	6.54
Selenium	Se	34	78.96	2–	6,8	9.75
Silicon	Si	14	28.09	4+	4	8.15
Silver	Ag	47	107.87	1+	4,6,8	7.57
Sodium	Na	11	22.99	1+	6,8	5.14
Strontium	Sr	38	87.62	2+	6,8,12	5.69
Sulfur	S	16	32.07	2–,6+	4,6,8	10.36
Tantalum	Ta	73	180.95	5+	6,8	7.88
Technetium	Tc	43	98.91	4+	6	7.28
Tellurium	Te	52	127.60	2–,4+	3	9.01
Terbium	Tb	65	158.93	3+	6,8	5.98
Thallium	Tl	81	204.38	1+	6,8	6.11
Thorium	Th	90	232.04	4+	6,8	6.95
Thulium	Tm	69	168.93	3+	6,8	5.81
Tin	Sn	50	118.71	2+	8	7.34
Titanium	Ti	22	47.88	2+,3+,4+	6	6.82
Tungsten (see Wolfram)						
Uranium	U	92	238.03	4+,6+	8,6,4	6.08

Appendix I. Atomic Parameters

Element	Symbol	Atomic Number	Atomic Weight[1], $^{12}C = 12.000$	Usual Valence[2]	Usual Coordination[2]	First Ionization[2] Potential, eV
Vanadium	V	23	50.94	5+	4,6	6.74
Wolfram	W	74	183.85	4+,6+	6,4	7.98
Xenon	Xe	54	131.29	0	0	12.13
Ytterbium	Yb	70	173.04	3+	6,8	6.2
Yttrium	Y	39	88.91	3+	6,8	6.38
Zinc	Zn	30	65.39	2+	4,6	9.39
Zirconium	Zr	40	91.22	4+	6,8	6.84

[1]International Union of Pure and Applied Chemistry 1986. Atomic weights of the elements 1985. *Pure and Apl. Chem.* 58 1677–1692. © 1986 IUPAC.

[2]Elements having closely similar values of valence (ionic charge), coordination (number of nearest neighbors), and first ionization potential may substitute for each other in mineral structures.

Appendix II
Mineral Determinative Tables

The inherent difficulties of mineral identification may be reduced by grouping together minerals that have closely similar physical characteristics in order to limit the number of minerals to choose between. This procedure will lead to small groups of minerals, and final identification may then be made by a simple chemical test, some distinctive property, the mineral occurrence, or by recognition of associated minerals.

For a number of reasons, it is difficult both to construct and to use determinative tables on a completely logical and objective basis. Mineral determination is, to a large degree, accomplished by a subjective summation of all of the mineral characteristics and is ordinarily not a step-by-step process. Physical properties show relatively large variations and, unless the mineral under study is a completely average specimen, its classification may be impossible without recourse to chemical, optical, or x-ray testing. Small grain size, incipient alteration, pseudomorphism, staining, and many other phenomena make the determination of critical properties difficult or impossible. It is recommended that many small mineral grains be separated from their matrix and examined against a white paper background with a good lens or binocular microscope in order to judge luster, cleavage, and color. Hardness may be tested on fine-grained material in matrix by the use of a needle.

Two listings and a set of determinative tables are given below as aids to the determination of minerals. Table I lists those ore and gangue minerals that may be found in some of the more common geologic environments; in general, the minerals in any column may be found together. Table II groups those minerals that display some relatively striking physical properties. Table III is a determinative table in which minerals are grouped into sections according to their luster, cleavage, hardness, and color. The table is divided into two principal parts in which minerals with metallic and nonmetallic lusters are listed. Each of these parts is further divided into sections based on cleavage and color in accordance with the following outline:

PART I

Minerals with a Metallic Luster
(Opaque on thinnest edge and having a dark-colored streak)

Section A: No cleavage
Section B: Distinct cleavage

PART II

Minerals with a Nonmetallic Luster
(Transparent on thinnest edge and having a colorless or light-colored streak)

Appendix II. Mineral Determinative Tables

Section A: No cleavage
Section B: One cleavage yielding tabular or micaceous fragments
Section C: Cleavage fragments are polyhedral
Section D: Distinct cleavage, but not yielding fragments classified in Sections B and C above
    Colorless, white, or lightly tinted minerals
Section E: Distinct cleavage, but not yielding fragments classified in Sections B and C above
    Black, brown, or red minerals
Section F: Distinct cleavage, but not yielding fragments classified in Sections B and C above
    Orange, yellow, green, blue, or violet minerals

The minerals in each section are arranged in the order of their increasing hardness, which is shown graphically in the left-hand column.

The mineral formulae will provide clues to critical chemical spot tests. Specific gravity may be a useful criterion, especially if its value has been determined or the contrast of specific gravity of two possibilities is large. Color, and especially streak—the color of the powdered mineral—may be essential clues in the identification of a particular species, but color must be used with caution. Particular features of determinative value are given in the remarks column.

Multiple entries of minerals have been made for those species in which confusion might arise because the critical parameters are difficult to assess or because the variations in physical properties are large enough to warrant several listings. For example, sphalerite will be found under both metallic and nonmetallic luster, gypsum in Sections B, C, and D of Part II, and orthoclase under both uncolored and red cleavable minerals.

The use of the tables may be illustrated by a few examples:

1. OBSERVATIONS: The mineral is nonmetallic, blue-green, shows cleavage, is softer than 3, and is water soluble.
   DETERMINATION: The mineral will be found in Part II, Section F. The only minerals fitting the observations are melanterite and chalcanthite. A test for copper or iron will readily distinguish these two species.
2. OBSERVATIONS: The mineral has a metallic luster, no cleavage, hardness between 3 and 5.5, yellowish color, and black streak.
   DETERMINATION: The mineral will be found between romanechite and magnetite, inclusive, in Part I, Section A. The mineral is gold, chalcopyrite, or pyrrhotite, based on color, and chalcopyrite or pyrrhotite based on streak. These two minerals can be readily distinguished by a test for copper (chalcopyrite, if positive) or magnetism (pyrrhotite, if positive).
3. OBSERVATIONS: The mineral is nonmetallic, has a micaceous cleavage yielding flexible folia, can be scratched by a copper penny, and is sectile. Further, the mineral does not have an unctuous or greasy feel nor an earthy smell when wet.
   DETERMINATION: The mineral will be found between the smectite group and barite, inclusive, in Part II, Section B. Color and streak somewhat limit the choice, and feel and smell further reduce the possibilities. The mineral must

be gypsum, muscovite, lepidolite, or barite. The occurrence of the mineral in a bed with halite together with the evolution of copious water when heated, low specific gravity, and sectility identify the mineral as gypsum.

4. OBSERVATIONS: The mineral is metallic and shows distinct cleavage. The hardness is about 3, and the color and streak are black.
   DETERMINATION: The mineral will be found in Part I. Section B. The black color and streak coupled with the hardness make jamesonite, bournonite, and enargite the only possibilities. The distinctive habits of these minerals may serve to distinguish them, or combinations of tests for Cu, Pb, and Fe can be used.

5. OBSERVATIONS: The mineral is nonmetallic, yields rhombic cleavage fragments, has a hardness of about 3.5, and is pink in color.
   DETERMINATION: The mineral will be found in Part II, Section C. Only one pink mineral, rhodochrosite, is listed, but a confirmatory test with cold, dilute HCl is negative because strong effervescence is observed. The mineral must therefore be a carbonate, and only calcite satisfies the observations.

Table A2.1
Occurrence of Ore and Gangue Minerals

Igneous Rocks			Metamorphic Rocks		
Silica-rich	Silica-poor	Mafic Volcanics	Crystalline Schists and Gneisses	Calcareous Metamorphic Rocks	Contact Metamorphic Rocks
Molybdenite	Platinum	Copper	Graphite	Graphite	Sphalerite
		Diamond			Chalcopyrite
Fluorite	Chalcopyrite		Pyrrhotite	Arsenopyrite	Arsenopyrite
	Pyrrhotite	Anhydrite	Arsenopyrite	Calcite	Molybdenite
Cassiterite	Nickeline			Dolomite	
Magnetite	Pentlandite	Chalcedony	Rutile		Corundum
		Opal	Corundum	Phlogopite	Hematite
Monazite	Corundum	Plagioclase		Olivine	Cassiterite
Apatite	Ilmenite	Prehnite	Rhodocrosite	Tremolite	Magnetite
	Rutile	Datolite		Garnet	
Quartz	Magnetite		Monazite	Epidote	Scheelite
Microcline	Chromite		Apatite		
Orthoclase			Pyromorphite		Tremolite
Plagioclase	Apatite				Diopside
Muscovite			Quartz		Tourmaline
Garnet	Plagioclase		Orthoclase		Vesuvianite
Hornblende	Nepheline		Plagioclase		
Zircon	Chrysotile		Chlorite		
Titanite	Enstatite		Muscovite		
	Diopside		Biotite		
	Augite		Hornblende		
	Olivine		Diopside		
	Garnet		Vesuvianite		
	Titanite		Garnet		
			Zircon		

Appendix II. Mineral Determinative Tables

Table A2.1
Continued

Sedimentary Rocks			Veins	Oxidized Zones	Volcanic Sublimates
Sulfur	Apatite	Gold	Barite	Copper	Sulfur
	Collophane	Silver	Celestine		
Bornite	Carnotite	Arsenic	Anhydrite	Chalcocite	Realgar
Chalcopyrite			Jarosite	Bornite	Orpiment
Marcasite	Quartz	All sulfides		Covellite	
	Microcline		Apatite	Pyrargyrite	Halite
Halite	Kaolinite	Sulfosalts			Fluorite
Sylvite	Smectite		Wolframite	Chlorargyrite	
Fluorite	Muscovite	Fluorite	Scheelite		Anhydrite
Carnallite	Glauconite		Quartz	Cuprite	Gypsum
		Hematite	Microcline	Hematite	
Hematite		Ilmenite	Rhodonite	Cassiterite	
Uraninite		Rutile	Topaz	Romanechite	
		Cassiterite			
Gibbsite		Uraninite		Smithsonite	
Diaspore		Magnetite		Aragonite	
Bauxite				Malachite	
		Brucite		Azurite	
Magnesite		Manganite			
Siderite					
Aragonite		Calcite		Anglesite	
Strontianite		Magnesite		Gypsum	
		Siderite		Melanterite	
Borax		Rhodochrosite		Epsomite	
Colemanite		Witherite		Pyromorphite	
		Strontianite		Vanadinite	
Barite		Dolomite			
Celestine		Cerussite		Wulfenite	
Anhydrite				Chrysocolla	
Melanterite				Hemimorphite	
Polyhalite					

Table A2.2
Minerals Showing Some Distinctive Physical Properties

Sectile Minerals	Magnetic Minerals	Radioactive Minerals	Luminescent or Fluorescent Minerals	Piezoelectric, Pyroelectric, Thermoelectric, and Triboelectric Minerals	Easily Fused Minerals
Gold	Platinum	Uraninite	Diamond	Diamond	Sulfur
Silver				Graphite	
Platinum	Pyrrhotite	Monazite	Sphalerite		Realgar
Graphite		Carnotite			Orpiment
Sulfur	Ilmenite[1]	Autunite	Fluorite	Arsenopyrite	Stibnite
	Hematite[1]	Tobernite		Sphalerite	Bismuthinite
Acanthite	Magnetite		Corundum		Skutterudite
Chalcocite[2]	Chromite	Zircon		Alunite	
Cinnabar		Titanite	Calcite	Jarosite	Pyrargyrite
Realgar	Wolframite		Aragonite	Pyromorphite	Proustite
Orpiment			Witherite		Enargite
Stibnite[2]			Strontianite	Quartz	Bournonite
Bismuthinite[2]			Cerussite	Tourmaline	
Molybdenite			Dolomite	Hemimorphite	Chlorargyrite
			Magnesite		
Chlorargyrite					Ice
			Celestine		
Wulfenite[2]			Anglesite		
Pyrophyllite			Apatite		
Talc					
Chrysocolla[2]			Scheelite		
Garnierite[2]					
			Quartz		
			Wollastonite		

Minerals Whose Color May Change on Heating	Water-soluble Minerals	Minerals Whose Specific Gravity > 3.5 but < 5.0		Minerals Whose Specific Gravity > 5.0	
Rutile	Sylvite	Diamond	Witherite	Gold	Cuprite
Corundum	Halite		Strontianite	Silver	Hematite
Spinel		Sphalerite	Malachite	Platinum	Pyrolusite
	Borax	Chalcopyrite	Azurite	Arsenic	Cassiterite
Apatite	Kernite	Pyrrhotite			Uraninite
Turquoise		Covellite	Barite	Acanthite	Magnetite
	Mirabilite		Celestine	Chalcocite	Columbite-Tantalite
Vivianite		Realgar			Series
Crocoite	Melanterite	Orpiment	Carnotite	Bornite	
	Epsomite	Stibnite		Galena	
Sodalite	Chalcanthite	Marcasite	Wolframite	Niccoline	Cerussite

Table A2.2
Continued

Minerals Whose Color May Change on Heating	Water-soluble Minerals	Minerals Whose Specific Gravity > 3.5 but < 5.0		Minerals Whose Specific Gravity > 5.0	
Muscovite		Molybdenite		Cinnabar	
Biotite		Pentlandite	Enstatite	Pyrite	Anglesite
Rhodonite		Enargite	Diopside	Cobaltite	Monazite
Beryl			Jadeite	Arsenopyrite	Pyromorphite
Olivine		Corundum	Rhodonite	Skutterudite	Vanadinite
Zircon		Ilmenite	Hemimorphite	Bismuthinite	Carnotite
Topaz		Spinel	Garnet		
Titanite		Chromite	Zircon	Pyrargyrite	Scheelite
			Epidote	Proustite	Wulfenite
		Goethite	Topaz	Tetrahedrite	Crocoite
		Limonite	Titanite	Bournonite	
		Romanechite		Jamesonite	
		Siderite		Chlorargyrite	
		Rhodochrosite			
		Smithsonite			

[1] Included magnetite
[2] Subsectile

Table A2.3.
Part I: Metallic Luster, Section A: No Cleavage

Hardness 1 2 3 4 5 6 7 8 9 10	Hardness Range	Mineral Name	Crystal System	Formula
—	2–2.5	Acanthite	Mon	Ag_2S
————	2–6.5	Pyrolusite	Tet	MnO_2
—	2.5–3	Gold	Isom	Au
—	2.5–3	Silver	Isom	Ag
—	2.5–3	Copper	Isom	Cu
—	2.5–3	Chalcocite	Mon	Cu_2S
—	2.5–3	Bournonite	Orth	$PbCuSbS_3$
·	3	Bornite	Tet	Cu_5FeS_4
—	3–4.5	Tetrahedrite Series	Isom	$(Cu,Fe)_{12}(Sb,As)_4S_{13}$
—	3.5–4	Pentlandite	Isom	$(Ni,Fe)_9S_8$
—	3.5–4	Chalcopyrite	Tet	$CuFeS_2$
—	3.5–4.5	Pyrrhotite	Mon + hex	$Fe_{1-x}S$
—	4–4.5	Platinum	Isom	Pt
—	5–5.5	Nickeline	Hex	NiAs
—	5–6	Romanechite	Mon	$BaMnMn_8O_{16}(OH)_4$
—	5–6.5	Hematite	Hex	Fe_2O_3
—	5–6	Ilmenite Series	Hex	$(Fe,Mg,Mn)TiO_3$
—	5–6	Uraninite	Isom	UO_2
—	5.5	Chromite	Isom	$FeCr_2O_4$
—	5.5–6.5	Magnetite	Isom	$FeFe_2O_4$
—	5.5–6.5	Franklinite	Isom	$(Fe,Zn,Mn)Fe_2O_4$
—	6–6.5	Pyrite	Isom	FeS_2
—	6–6.5	Marcasite	Orth	FeS_2

Quartz
Knife blade
Copper penny

Appendix II. Mineral Determinative Tables

Specific Gravity	Color	Streak	Remarks
7.3	Dark lead-gray	Black and shining	Sectile, darkens on exposure to light
5.0	Steel-gray	Black	Pulverulent forms may soil fingers, crystals hard
19.3	Yellow	Yellow	Sectile, does not tarnish
10.5	Silver-white	Silver-white	Sectile, black tarnish
8.9	Copper-red	Copper-red	Sectile, brown tarnish
5.7	Dark lead-gray	Dark lead-gray	Subsectile, massive
5.8	Gray to black	Gray to black	Wheellike twins, easily fused
5.1	Copper-red	Gray-black	"Peacock" tarnish
4.4–5.4	Gray to black	Black to red	Very thin splinters are red in transmitted light
4.6–5	Light bronze-yellow	Light bronze-brown	Brittle, with pyrrhotite in mafic rocks
4.2	Brass-yellow	Greenish black	Common sulfide
4.6	Bronze-yellow	Black	Weakly magnetic
14–19	Gray	Gray	Sectile, may be weakly magnetic, in ultramafic rocks
7.3–7.7	Pale copper-red	Brownish black	Green alteration "bloom"
4.7	Black to gray	Black and shining	Colloform or earthy
5.3	Red-brown to black	Brick-red	Common mineral
4.7	Black	Black	Common mineral, may be weakly magnetic
10.8 or less	Black	Brownish black	Radioactive, usually massive or colloform
4.6	Black	Brown	May be weakly magnetic, in mafic rocks
5.2	Black	Black	Common mineral, strongly magnetic
5.1–5.2	Black	Brown	With willemite and zincite, weakly magnetic
5.0	Pale brass-yellow	Green-black to brown-black	Common mineral, crystals as cubes or pyritohedra
4.9	White to pale yellow	Gray-black	Tarnishes to brass-yellow, "coxcomb" crystals

Table A2.3.
Part I: Metallic Luster, Section B: Distinct Cleavage

Hardness 1 2 3 4 5 6 7 8 9 10	Hardness Range	Mineral Name	Crystal System	Formula
	1–1.5	Molybdenite	Hex	MoS_2
	1–2	Graphite	Hex	C
	1.5–2	Covellite	Hex	CuS
	2	Stibnite	Orth	Sb_2S_3
	2–2.5	Acanthite	Mon	Ag_2S
	2–2.5	Cinnabar	Hex	HgS
	2.5	Galena	Isom	PbS
	2.5	Pyrargyrite	Hex	Ag_3SbS_3
	2.5	Jamesonite	Mon	$Pb_4FeSb_6S_{14}$
	2.5–3	Bournonite	Orth	$PbCuSbS_3$
	3	Bornite	Tet	Cu_5FeS_4
	3–3.5	Enargite	Orth	Cu_3AsS_4
	3.5	Arsenic	Hex	As
	3.5–4	Sphalerite	Isom	ZnS
	3.5–4	Pentlandite	Isom	$(Ni,Fe)_9S_8$
	3.5–4	Cuprite	Isom	Cu_2O
	3.5–4.5	Pyrrhotite	Mon + hex	$Fe_{1-x}S$
	4	Manganite	Mon	MnO(OH)
	4–4.5	Wolframite Series	Mon	$(Fe,Mn)(WO_4)$
	5–5.5	Goethite	Orth	FeO(OH)
	5–6	Ilmenite Series	Hex	$(Fe,Mg,Mn)TiO_3$
	5.5	Cobaltite Group	Orth	CoAsS
	5.5–6	Arsenopyrite	Mon	FeAsS
	5.5–6	Skutterudite Series	Isom	$(Co,Ni,Fe)As_{3-x}$
	5.5–6.5	Magnetite	Isom	$FeFe_2O_4$
	5.5–6.5	Franklinite	Isom	$(Fe,Zn,Mn)Fe_2O_4$
	6–6.5	Rutile	Tet	TiO_2
	6–6.5	Columbite-Tantalite Series	Orth	$(Fe,Mn)(Nb,Ta)_2O_6$
	6–6.5	Pyrolusite	Tet	MnO_2
	6–7	Cassiterite	Tet	SnO_2

Quartz
Knife blade
Copper penny

Specific Gravity	Color	Streak	Remarks
4.7	Gray	Gray to black	Foliated, distinguished from graphite by green streak on glazed porcelain
2.1–2.3	Steel-gray to black	Black	Foliated, greasy feel
4.7	Indigo-blue	Black to gray	Platy aggregates
4.6	Gray	Gray	Columnar, black tarnish
7.3	Dark lead-gray	Black and shining	Sectile
8.1	Red	Red	Subsectile, often earthy
7.6	Lead-gray	Lead-gray	Cubic crystals and cleavage
5.8	Deep red	Red	Translucent
5.6	Gray-black	Gray-black	Plumose, iridescent tarnish
5.8	Gray to black	Gray to black	Wheel-like twins
5.1	Copper-red	Gray-black	"Peacock" tarnish
4.4	Black	Black	Bladed aggregates
5.7	Tin-white	Gray	Dark gray tarnish, concentric layers
3.9–4.2	Black to brown	Brown to yellow	Common sulfide, resinous luster
4.6–5.0	Light bronze-yellow	Yellow-brown	Octahedral parting, with pyrrhotite and chalcopyrite in mafic rocks
6.1	Dark red	Brown-red	Highly lustrous octahedra
4.6	Bronze-yellow	Black	Weakly magnetic
4.2–4.4	Black	Brown	Prismatic crystals, with other Mn oxides
7.1–7.5	Black to brown-black	Black to brown-black	Sometimes weakly magnetic
3.3–4.3	Brown	Brownish yellow	Common mineral, colloform
4.5–5.0	Black to red	Black to reddish to yellow	Sometimes weakly magnetic, massive
6.3	Silver-white	Gray-black	Pink alteration "bloom"
6.1	Silver-white	Black	Prismatic crystals, common
6.1–6.9	Tin-white to silver-gray	Black	Cubic or octahedral crystals
5.2	Black	Black	Common mineral, strongly magnetic
5.1–5.2	Black	Brown	With willemite and zincite, weakly magnetic
4.2	Red to black	Pale brown	Prismatic crystals
5.2–7.9	Black to brown-black	Dark red to black	In granitic pegmatites
5.0	Steel-gray	Black	Pulverulent forms may stain fingers, crystals hard
6.8–7.1	Brown or black	White to brown	In greisen, highly lustrous

Table A2.3.
Part II: Nonmetallic Luster, Section A: No Cleavage

Hardness 1 2 3 4 5 6 7 8 9 10	Hardness Range	Mineral Name	Crystal System	Formula
-	1	Carnallite	Orth	$KMgCl_3 \cdot 6H_2O$
━━	1–3	Bauxite	—	Mixture of Al hydroxides
▬	1–1.5	Limonite	—	$Fe_2O_3 \cdot nH_2O$
▬	2–2.5	Cinnabar	Hex	HgS
━	2–3	Chlorargyrite	Isom	$AgCl$
━	2–3	Garnierite	—	$(Ni,Mg)_3(Si_2O_5)(OH)_4$
━━	2–4	Chrysocolla	—	$(Cu,Al)(SiO_3) \cdot nH_2O$
━━━	2–5	Collophane	Mon	$Ca_3(PO_4)_3 \cdot H_2O$
▬	2.5–3.5	Serpentine Group	Mon	$(Mg,Fe)_3(Si_2O_5)(OH)_4$
-	3	Vanadinite	Hex	$Pb_5(VO_4)_3Cl$
▬	3–4	Pyromorphite Series	Hex	$Pb_5[(PO_4),(AsO_4),(VO_4)]Cl$
▬	5	Apatite Series	Hex	$Ca_5(PO_4)_3(F,Cl,OH)$
━	5–6	Uraninite	Isom	UO_2
━	5–6	Turquoise	Tric	$CuAl_6(PO_4)_4(OH)_8 \cdot 5H_2O$
━	5–6	Opal	—	$SiO_2 \cdot nH_2O$
━	5–6.5	Hematite	Hex	Fe_2O_3
▬	6–7	Cassiterite	Tet	SnO_2
-	6.5	Vesuvianite	Tet	$Ca_{10}(Mg,Fe)_2Al_4(Si_2O_7)_2(SiO_4)_5(OH,F)_4$
▬	6.5–7	Olivine Series	Orth	$(Mg,Fe,Mn)_2(SiO_4)$
━	6.5–7.5	Garnet Group	Isom	$X_3^{2+}Y_2^{3+}(SiO_4)_3$
-	7	Chalcedony	Hex	SiO_2
-	7	Quartz	Hex	SiO_2
▬	7–7.5	Tourmaline Series	Hex	$WX_3Y_6(BO_3)_3(Si_6O_{18})(OH,F)_4$
▬	7.5–8	Spinel	Isom	$MgAl_2O_4$
▬	7.5–8	Beryl	Hex	$Be_3Al_2(Si_6O_{18})$
-	9	Corundum	Hex	Al_2O_3

Quartz
Knife blade
Copper penny

Specific Gravity	Color	Streak	Remarks
1.6	White to reddish	White	In salt beds, bitter taste
2.0–2.6	White, stained gray, yellow, or red	White	Often pisolitic, Al-ore
3.6–4.0	Brown to black	Yellow-brown	Alteration of iron-bearing minerals
8.1	Dark red	Red	Subsectile, earthy
5.5	Gray to green	White	Texture of horn, color deepens on exposure to light
2.5	Green to white	White	Unctuous, earthy
2.3	Green to brown	White	Encrustations and fillings, enamel-like
2.6–2.9	White, gray, or brown	White	Colloform variety of apatite
2.2–2.4	Green	White	Smooth or greasy feel, common alteration product
6.7–7.1	Red to yellow	White to yellow	Hollow prisms
6.5–7.2	Green, brown, or yellow	White	Hexagonal prisms, often barrel-shaped
3.1–3.4	Green or red	White	Hexagonal prisms
10.8 or less	Black	Brownish black	Radioactive
2.6–2.8	Blue to green	White or greenish	Turns brown when heated
1.9–2.2	Various	White	Milky or opalescent, never in crystals
5.3	Red-brown to black	Brick-red	Common mineral
6.8–7.1	Brown or black	White to brown	Highly lustrous, in greisen
3.4	Brown to green	White	Columnar aggregates, in metamorphic rocks
3.3–4.4	Green to brown	White	Rock-forming mineral, in mafic rocks
3.5–4.3	Red, brown, or green	White	Dodecahedra or trapezohedra
2.6	Various	White	Often banded
2.6	Colorless or white	White	Common rock-forming mineral, glassy
3.0–3.2	Black, green, or red	White	Spherical triangle cross-section
3.6	Various	White	Octahedra
2.6–2.9	Blue to green	White	Hexagonal prisms in pegmatite
4.0	Brown, gray, pink, or blue	White	Good parting, in metamorphic rocks

Table A2.3.
Part II: Nonmetallic Luster, Section B: Distinct Cleavage
(Cleavage Fragments Are Tabular or Micaceous.)

Hardness 1 2 3 4 5 6 7 8 9 10	Hardness Range	Mineral Name	Crystal System	Formula
—	1	Smectite Group	Mon	$X(Al,Mg)_2(Si_4O_{10})(OH)_2 \cdot nH_2O$
	1–1.5	Talc	Mon	$Mg_3(Si_4O_{10})(OH)_2$
	1–2	Pyrophyllite	Mon	$Al_2(Si_4O_{10})(OH)_2$
—	1–2	Graphite	Hex	C
▬	1–2	Carnotite	Mon	$K_2(UO_2)_2(VO_4)_2 \cdot 3H_2O$
•	1.5	Vermiculite	Mon	$Mg_3(Si_4O_{10})(OH)_2 \cdot nH_2O$
—	1.5–2	Orpiment	Mon	As_2S_3
▬	2	Gypsum	Mon	$Ca(SO_4) \cdot 2H_2O$
—	2–2.5	Kaolinite	Tric	$Al_2(Si_2O_5)(OH)_4$
—	2–2.5	Chlorite Group	Mon	$Y_n(Z_4O_{10})(OH)_8$
—	2–2.5	Autunite	Tet	$Ca(UO_2)_2(PO_4)_2 \cdot nH_2O$
—	2.5–3	Muscovite	Mon	$KAl_2(AlSi_3O_{10})(OH,F)_2$
—	2.5–3	Biotite	Mon	$K(Mg,Fe)_3(AlSi_3O_{10})(OH)_2$
—	2.5–3	Phlogopite	Mon	$KMg_3(AlSi_3O_{10})(OH,F)_2$
—	2.5–3.5	Gibbsite	Mon	$Al(OH)_3$
▬	2.5–3.5	Jarosite	Hex	$KFe_3(SO_4)_2(OH)_6$
▬	2.5–4	Lepidolite Series	Mon	$KX_3(Al,Si)_4O_{10}(OH,F)_2$
•	3	Wulfenite	Tet	$Pb(MoO_4)$
—	3–3.5	Barite	Orth	$Ba(SO_4)$
•	3.5	Arsenic	Hex	As
—	6–6.5	Albite	Tric	$Na(AlSi_3O_8)$

Quartz
Knife blade
Copper penny

Appendix II. Mineral Determinative Tables

Specific Gravity	Color	Streak	Remarks
2.0–7	White or gray	White	Unctuous feel, earthy smell when wet
2.7	Green to white	White	Greasy feel, plastic cleavage folia
2.8	White	White	Greasy feel, plastic cleavage folia
2.1–2.3	Steel-gray to black	Black	Greasy feel
4–5	Yellow	Pale yellow	Radioactive, in sandstones
2.4	Yellow to brown	White	Expands when heated
3.5	Yellow	Yellow	Sectile, melts at 300° C
2.3	White, gray, or brown	White	Common mineral, plastic
2.6	White, often stained	White	Unctuous when wet, earthy
2.7	Dark green	White	Common alteration product, flexible cleavage folia
3.1	Yellow	Yellow	Tabular crystals, radioactive
2.8–3.0	Colorless or tints	White	Common rock-forming mineral, elastic cleavage folia
2.7–3.1	Black or brown-black	White	Common rock-forming mineral, elastic cleavage folia
2.8	Brown	White	Rare in igneous rocks, elastic cleavage folia
2.4	White, often stained	White	Earthy smell when wet, in bauxite deposits and talc schists
2.9–3.3	Yellow to brown	Pale yellow	Crusts and coatings
2.9	Pink, violet, or white	White	Elastic cleavage folia, in pegmatites
6.5–7.0	Orange to yellow	White	Square tablets
4.5	White	White	Often tinted yellow, red, green, or blue; tabular crystals
5.7	Tin-white	Gray	Dark gray tarnish, concentric layers
2.6	White	White	A plagioclase

Table A2.3.
Part II: Nonmetallic Luster, Section C: Distinct Cleavage (Cleavage Fragments Are Polyhedral.)

Hardness 1 2 3 4 5 6 7 8 9 10	Hardness Range	Mineral Name	Crystal System	Formula
—	2	Gypsum	Mon	$Ca(SO_4) \cdot 2H_2O$
—	2	Halite	Isom	$NaCl$
—	2	Sylvite	Isom	KCl
—	3–3.5	Calcite	Hex	$Ca(CO_3)$
—	3–3.5	Celestine	Orth	$Sr(SO_4)$
·	3.5	Anhydrite	Orth	$Ca(SO_4)$
—	3.5–4	Sphalerite	Isom	ZnS
—	3.5–4	Rhodochrosite	Hex	$Mn(CO_3)$
—	3.5–4	Dolomite	Hex	$CaMg(CO_3)_2$
—	3.5–4	Cuprite	Isom	Cu_2O
—	4–4.5	Siderite	Hex	$Fe(CO_3)$
—	4	Magnesite	Hex	$Mg(CO_3)$
—	4	Fluorite	Isom	CaF_2
—	4–4.5	Smithsonite	Hex	$Zn(CO_3)$
—	10	Diamond	Isom	C

Quartz
Knife blade
Copper penny

Appendix II. Mineral Determinative Tables

Specific Gravity	Color	Streak	Remarks
2.3	White, gray, or brown	White	Flattened rhombohedra, fibrous, massive
2.2	Colorless or white	White	Cubes, salty taste
2.0	White to bluish or reddish	White	Cubes, bitter salty taste, often tinted
2.7	Colorless or white	White	Rhombs, often colored, very common
4.0	Colorless to pale blue	White	Oblongs, in sedimentary rocks
3.0	Colorless to blue-white	White	Oblongs, often tinted
3.9–4.2	Yellow, brown, black	White, Yellow	Dodecahedra, resinous luster
3.3–3.7	Pink	White	Rhombs
2.9	White	White	Rhombs, common sedimentary rock-forming mineral
6.1	Red	Brown-red	Octahedra, highly lustrous
3.9	Brown	White	Rhombs
3.0	Colorless or white	White	Rhombs or white enamel as an alteration of Mg-rich rock
3.2	Violet, green, white, or blue	White	Octahedral cleavage, cubic crystals, often fluorescent
4.4	White, green	White	Rhombs, usually colloform, variously colored
3.5	Colorless, bluish, pink, green, yellow, or black	None	Octahedra

Table A2.3.
Part II: Nonmetallic Luster, Section D: Distinct Cleavage (Colorless or Lightly Tinted Minerals; Cleavage Fragments Are Not Micaceous, Tabular, or Polyhedral.)

Hardness (1 2 3 4 5 6 7 8 9 10)	Hardness Range	Mineral Name	Crystal System	Formula
— at 2	2	Gypsum	Mon	$Ca(SO_4) \cdot 2H_2O$
— at 2–2.5	2–2.5	Borax	Mon	$Na_2(B_4O_7) \cdot 10H_2O$
—— at 2.5–3	2.5–3	Anglesite	Orth	$Pb(SO_4)$
— at 3–3.5	3–3.5	Witherite	Orth	$Ba(CO_3)$
— at 3–3.5	3–3.5	Cerussite	Orth	$Pb(CO_3)$
— at 3–3.5	3–3.5	Celestine	Orth	$Sr(SO_4)$
— at 3–3.5	3–3.5	Barite	Orth	$Ba(SO_4)$
— at 3.5–4	3.5–4	Strontianite	Orth	$Sr(CO_3)$
— at 3.5–4	3.5–4	Aragonite	Orth	$Ca(CO_3)$
· at 4	4	Alunite	Hex	$KAl_3(SO_4)_2(OH)_6$
— at 4.5–5	4.5–5	Scheelite	Tet	$Ca(WO_4)$
—— at 4.5,6.5	4.5, 6.5	Kyanite	Tric	$AlAl(SiO_4)O$
— at 5–6	5–6	Nepheline	Hex	$(KNa)(AlSiO_4)$
— at 5–6	5–6	Tremolite Series	Mon	$X_2Y_5(Si_8O_{22})(OH)_2$
—— at 5.5–6.5	5.5–6.5	Diopside Series	Mon	$Ca(Mg,Fe)(Si_2O_6)$
· at 6	6	Adularia	Mon	$(KNa)(AlSi_3O_8)$
— at 6–6.5	6–6.5	Orthoclase Series	Mon	$(K,Na)(AlSi_3O_8)$
— at 6–6.5	6–6.5	Microcline Series	Tric	$(K,Na)(AlSi_3O_8)$

Specific Gravity	Color	Streak	Remarks
2.3	Colorless to white, gray, brown	White	Common mineral, plastic
1.7	Colorless to white	White	Sweetish taste, in evaporites, tinted gray, blue, or green
6.4	Colorless to white	White	Alteration of galena, tinted gray, yellow, brown, or green
4.3	Colorless to white	White	Tinted yellow, brown, or green
6.6	Colorless to white	White	Reticulated aggregates; gray, blue, or green tints; very brittle
4.0	Colorless to pale blue	White	In sedimentary rocks, tinted red or green
4.5	Colorless to white	White	Tinted yellow, brown, red, gray, green, or blue; tabular crystals. Heavy.
3.8	Colorless to gray	White	Columnar or fibrous; tinted yellow, green, or red
2.9	Colorless to white	White	Columnar or fibrous
2.6–2.8	White to gray or pink	White	Alteration of salic volcanics
6.1	Colorless to white and gray	White	Blue fluorescence; tinted yellow, green, or red
3.6	Blue-white	White	Bladed crystals in metamorphic rocks
2.6	Colorless to white	White	In silica-poor igneous rocks; tinted yellow, green, gray, or red
3.0–3.3	Gray to white	White	Prismatic crystals, in contact metamorphic zones
3.2–2.6	White to green	White	Stubby prismatic crystals
2.6	Colorless	White	Pseudo-orthorhombic crystals, often opalescent
2.6	White to flesh-pink	White	Common rock-forming mineral
2.6	White to cream to reddish, yellowish, or greenish	White	Common rock-forming mineral, distinguished from orthoclase by fine twinning striations

Table A2.3.
Part II: Nonmetallic Luster, Section D: Distinct Cleavage (Colorless or Lightly Tinted Minerals; Cleavage Fragments Are Not Micaceous, Tabular, or Polyhedral.)

Hardness 1 2 3 4 5 6 7 8 9 10	Hardness Range	Mineral Name	Crystal System	Formula
—	6–6.5	Plagioclase Series	Tric	$(Ca,Na)[(AlSi)_4O_8]$
—	6.5–7	Spodumene	Mon	$LiAl(Si_2O_6)$
—	7.5	Zircon	Tet	$Zr(SiO_4)$
—	7.5	Andalusite	Orth	$AlAl(SiO_4)O$
—	8	Topaz	Orth	$Al_2(SiO_4)(F,OH)_2$

Quartz
Knife blade
Copper penny

Table A2.3.
Part II: Nonmetallic Luster, Section E: Distinct Cleavage (Black, Brown, or Red Minerals; Cleavage Fragments Are Not Micaceous, Tabular, or Polyhedral.)

Hardness 1 2 3 4 5 6 7 8 9 10	Hardness Range	Mineral Name	Crystal System	Formula
—	1.5–2	Realgar	Mon	AsS
—	2–2.5	Proustite	Hex	Ag_3AsS_3
—	2.5	Pyrargyrite	Hex	Ag_3SbS_3
—	2.5–3	Crocoite	Mon	$Pb(CrO_4)$
—	2.5–3.5	Jarosite	Hex	$KFe_3(SO_4)_2(OH)_6$
—	3.5–4	Sphalerite	Isom	ZnS
—	3.5–4	Cuprite	Isom	Cu_2O
—	4	Manganite	Mon	$MnO(OH)$
—	4–4.5	Wolframite	Mon	$(Mn,Fe)(WO_4)$
—	5	Apatite Series	Hex	$Ca_5(PO_4)_3(F,Cl,OH)$
—	5–5.5	Goethite	Orth	$\alpha\text{-}FeO(OH)$

Appendix II. Mineral Determinative Tables

Specific Gravity	Color	Streak	Remarks
2.6–2.7	White or gray	White	Common rock-forming mineral, polysynthetic twinning
3.0–3.2	White	White	In pegmatites; tinted green, gray, yellow, purple, or pink
4.7	Brown to colorless	White	Square prisms; tinted gray, green, or red
3.2	White, rose, or green	White	Maltese cross on (001) section, in metamorphic rocks
3.5	Colorless to white	White	Highly modified prisms; tinted green, blue, or red

Specific Gravity	Color	Streak	Remarks
3.5	Red to orange-yellow	Orange to yellow	Melts at 310° C
5.6	Red	Red	With pyrargyrite
5.8	Deep red	Red	Translucent
5.9–6.1	Red	Orange	Secondary mineral in lead deposits
2.9–3.3	Brown to yellow	Pale yellow	Crusts and coatings
3.9–4.2	Black, brown, or yellow	Brown, yellow, or white	Resinous luster
6.1	Dark red	Brown-red	Octahedra, highly lustrous
4.3	Black	Brown	Prismatic crystals, with other Mn oxides
7.1–7.5	Black to brown-black	Brown-black	Sometimes weakly magnetic
3.1–3.4	Dark red-brown, blue, green	White	Hexagonal prisms
3.3–4.3	Brown	Brown	Common mineral, colloform

Table A2.3.
Continued

Part II: Nonmetallic Luster, Section E: Distinct Cleavage (Black, Brown, or Red Minerals; Cleavage Fragments Are Not Micaceous, Tabular, or Polyhedral.)

Hardness 1 2 3 4 5 6 7 8 9 10	Hardness Range	Mineral Name	Crystal System	Formula
—	5–5.5	Monazite	Mon	$(Ce,La,Nd,Th)(PO_4)$
—	5–5.5	Titanite	Mon	$CaTi(SiO_4)O$
—	5–6	Ilmenite Series	Hex	$(Fe,Mg,Mn)TiO_3$
—	5–6	Hornblende Series	Mon	$Ca_2(Fe,Mg)_4Al(AlSi_7O_{22})(OH)_2$
—	5–6	Enstatite Series	Orth	$(Mg,Fe)_2(Si_2O_6)$
—	5–6.5	Hematite	Hex	Fe_2O_3
—	5.5–6.5	Augite	Mon	$Ca(Mg,Fe,Al)[(Al,Si)_2O_6]$
—	5.5–6.5	Rhodonite	Tric	$Mn(SiO_3)$
—	6–6.5	Orthoclase Series	Mon	$(K,Na)(AlSi_3O_8)$
—	6–6.5	Microcline Series	Tric	$(K,Na)(AlSi_3O_8)$
—	6–6.5	Rutile	Tet	TiO_2
—	6–6.5	Columbite-Tantalite Series	Orth	$(Fe,Mn)(Nb,Ta)_2O_6$
—	6–7	Cassiterite	Tet	SnO_2
-	7.5	Zircon	Tet	$Zr(SiO_4)$
-	9	Corundum	Hex	$\alpha\text{-}Al_2O_3$

Quartz
Knife blade
Copper penny

Appendix II. Mineral Determinative Tables

Specific Gravity	Color	Streak	Remarks
5.1	Brown	White	Usually radioactive, accessory in salic igneous rocks
3.5	Brown, red, or black	White	Wedge-shaped crystals
4.5–5.0	Black, brown, red	Black, brown, or yellow	Common mineral, sometimes weakly magnetic
2.9–3.3	Black	White	Common rock-forming mineral
3.1–3.5	Brown	White	Common rock-forming mineral in mafic rocks
5.3	Black to red-brown	Brick-red	Common mineral, good parting
3.2–3.6	Black to dark green	White	Common rock-forming mineral, cleavage at right angles
3.4–3.7	Red or pink	White	Black surface stains
2.6	Red or pink	White	Common rock-forming mineral
2.6	Red to pink	White	Common rock-forming mineral, distinguished from orthoclase by fine twinning striations
4.2	Black to red	Pale brown	Prismatic crystals, highly lustrous
5.2–8.2	Black to brown-black	Dark red to black	In granitic pegmatites
7.0	Brown or black	White to brown	Highly lustrous, in greisen
4.7	Brown	White	Square prisms
4.0	Brown	White	In metamorphic rocks

Table A2.3.
Part II: Nonmetallic Luster, Section F: Distinct Cleavage (Orange, Yellow, Green, Blue, and Violet Minerals; Cleavage Fragments Are Not Micaceous, Tabular, or Polyhedral.)

Hardness 1 2 3 4 5 6 7 8 9 10	Hardness Range	Mineral Name	Crystal System	Formula
—	1.5–2	Orpiment	Mon	As_2S_3
—	1.5–2	Realgar	Mon	AsS
—	1.5–2.5	Sulfur	Orth	S
—	2	Melanterite	Mon	$Fe(SO_4) \cdot 7H_2O$
—	2.5	Chalcanthite	Tric	$Cu(SO_4) \cdot 5H_2O$
—	2.5–3.5	Jarosite	Hex	$KFe_3(SO_4)_2(OH)_6$
—	2.5–3.5	Serpentine Group	Mon	$Mg_3(Si_2O_5)(OH)_4$
—	2.5–5.5	Chrysotile Group	Mon	$(Mg,Fe)_3(Si_2O_5)(OH)_4$
—	3	Wulfenite	Tet	$Pb(MoO_4)$
—	3.5–4	Malachite	Mon	$Cu_2(CO_3)(OH)_2$
—	3.5–4	Azurite	Mon	$Cu_3(CO_3)_2(OH)_2$
—	4	Fluorite	Isom	CaF_2
— —	4.5, 6.5	Kyanite	Tric	$AlAl(SiO_4)O$
—	5	Apatite Series	Hex	$Ca_5(PO_4)_3(F,Cl,OH)$
—	5–5.5	Monazite	Mon	$(Ce,La,Nd,Th)(PO_4)$
—	5–6	Hornblende Series	Mon	$Ca_2(Fe,Mg)_4Al(AlSi_7O_{22})(OH)_2$
—	5–6	Enstatite Series	Orth	$(Mg,Fe)_2(Si_2O_6)$
—	5–6	Turquoise	Tric	$CuAl_6(PO_4)_4(OH)_8 \cdot 5H_2O$
—	5.5	Willemite	Hex	$Zn_2(SiO_4)$
—	5.5–6.5	Diopside Series	Mon	$Ca(Mg,Fe)_2(Si_2O_6)$
—	6–6.5	Prehnite	Orth	$Ca_2Al(AlSi_3O_{10})(OH)_2$
—	6–6.5	Microcline Series	Tric	$(K,Na)(AlSi_3O_8)$
—	6–7	Sillimanite	Orth	$AlAl(SiO_4)O$
—	6–7	Epidote Series	Mon	$X_2Y_3(SiO_4)(Si_2O_7)(O,OH)$
—	6.5–7	Jadeite	Mon	$NaAl(Si_2O_6)$
—	9	Corundum	Hex	$\alpha\text{-}Al_2O_3$

Quartz
Knife blade
Copper penny

Specific Gravity	Color	Streak	Remarks
3.5	Yellow	Yellow	Sectile, melts at 300° C
3.5	Orange-yellow to red	Orange-yellow to red	Sectile, melts at 310° C
2.1	Yellow	White	Melts and burns at 113° C
1.9	Blue or green	White	Sweetish taste
2.1–2.3	Blue	White	Metallic taste
2.9–3.3	Yellow to brown	Pale yellow	Crusts and coatings
2.2–2.4	Green	White	Smooth to greasy feel
2.2	Green	White	Asbestiform
6.5–7.0	Orange to yellow	White	Square tablets
4.1	Green	Green	Colloform
3.8	Blue	Blue	Encrusting
3.2	Green or violet	White	Cubic crystals with octahedral cleavage, frequently fluorescent
3.6	Blue-white	White	Bladed crystals in metamorphic rocks
3.1–3.4	Dark green	White	Hexagonal crystals
5.1	Yellow-brown	White	Usually radioactive
2.9–3.3	Dark green	White	Common rock-forming mineral
3.1–3.5	Greenish white	White	Common rock-forming mineral in mafic rocks
2.6–2.8	Blue to green	White or greenish	Turns brown when heated
3.9–4.2	Green to brown or white	White	With other zinc minerals, fluorescent
3.2–3.6	Green to white	White	Common mineral in mafic rocks and marbles
2.9	Light green	White	Colloform with a crystalline surface
2.6	White	White	In siliceous igneous rocks and pegmatites
3.2	Gray-green to white	White	Fibrous groups in metamorphic rocks
3.2–3.5	Green	White	Common mineral
3.3–3.5	Green	White	Compact and tough
4.0	Blue	White	Good parting, in metamorphic rocks

Appendix III

Market Specifications

Commodity	Form/Grade	Size	Basis	Price	Remarks
Alumina	calcined, varous grades	powder		xxx$/T	ton lots
Andalusite	mineral concentrate		cont. Al_2O_3	xx$/T	
Antimony	lump ore		60% cont. Sb	xx$/T	
	oxide		99.5% Sb_2O_3	xxx$/T	
	metal			x$/lb	
Aplite	glass grade	+200 mesh		xx$/T	
Arsenic	oxide	powder	95% As_2O_3	xx¢/lb	
	metal		99.5% As	xxx¢/lb	
Asbestos	chrysotile fibers				carlots
	Group 3 (spinning) and 4 (shingle)			xxxx$/T	
	Group 5 (paper), 6 (waste, stucco, plaster) and 7 (refuse or shorts)			xxx$/T	
Attapulgite				xxx$/T	ton lots
Barite	bulk			xx$/T	
	drilling mud grade	−350 mesh		xxx$/T	S.G. 4.2
	paint grade		96% $BaSO_4$	xxx$/T	ton lots
	micronized	−20 microns		xxx$/T	
Bastnasite	leached concentrate		cont. REO	x$/lb	
Bauxite	abrasive grade		86% Al_2O_3	xxx$/T	
	refractory grade		86% Al_2O_3	xxx$/T	
Bentonite	bulk			xx$/T	
Beryl	mineral concentrate		cont. BeO	xx$/T	
	oxide	powder	97% BeO	xx$/lb	
	metal			xxx$/lb	
Bismuth	metal			x$/lb	ton lots

Appendix III. Market Specifications

Commodity	Form Grade	Size	Basis	Price	Remarks
Boron	boric acid, various grades			xxx$/T	carlots
	borax, various grades			xxx$/T	carlots
	colemanite, crude	lump	44/46% B_2O_3	xxx$/T	carlots
	colemanite, calcined	−70 mesh	43% B_2O_3	xxx$/T	carlots
	ulexite	−7 mesh		xx$/T	carlots
Cadmium	metal			x$/lb	smelter byproduct
Celestie	ground		94% $SrSO_4$	xxx$/T	
Cesium	metal		99% Cs	xxx$/lb	
Chromite	refractory grade		35/36% Cr_2O_3	xx$/T	
	molding sand	−30 mesh		xx$/T	
	metallurgical grade		45% Cr_2O_3	xxx$/T	
	chemical grade		44/45% Cr_2O_3	xxx$/T	
Clay	ball clay				
	air dried, shredded			xx$/T	bulk lots
	refined, noodled			xx$/T	bulk lots
	pulverized, air floated			xxx$/T	
	flint clay, calcined			xxx$/T	
	kaolin				
	coating clays			xxx$/T	
	filler clays			xx$/T	
	pottery clays			xxx$/T	
Cobalt	metal		cont. Co	xx$/lb	
Columbium	pyrochlore concentrate		cont. Nb_2O_5	x¢/lb	
	columbite concentrate		cont. Nb_2O_5 + Ta_2O_5	xx$/slb	
	ferrocolumbium		63/68% Nb	x$/lb	
Copper	metal			xx–xxx¢/lb	
Diatomite	calcined			xxx$/T	
Emery	fine, medium, coarse	powder		xxx$/T	
Feldspar	ceramic grade	200 mesh		xx–xxx$/T	
	ceramic/glass grade	sand		xx$/T	
	glass grade	+200 mesh		xx$/T	
Fluorspar	metallurgical grade		70% CaF_2	xxx$/T	
	acid grade		97% CaF_2	xxx$/T	

Commodity	Form Grade	Size	Basis	Price	Remarks
	ceramic grade	powder		xxx$/T	
Gallium	metal		99.9999% Ga	xxx$/kg	
Gems			carat		
Garnet		graded powders		xxx$/T	ton lots
Germanium	metal			xxx$/kg	
	electronic grade GeO_2	powder		xxx$/kg	
Gold	refined, unfabricated		24 carat (1000 fine)	xxx$/oz	
Graphite		lump	90/99% C	xxxx$/T	
		chips	80/92% C	xxx$/T	
		dust	80/98% C	xxx$/T	
Gypsum	crude			x$/T	
	calcined			xx$/T	
Ilmenite	bulk concentrates		54/60% TiO_2	xx$/T	
	titanium slag			xxx$/T	
Indium	metal			xx$/oz	
Iron	natural ores	lump, fine	51.5% Fe	xx$/T	
	pellets			xx¢/T	long ton units
Iron oxide pigment	ochre	micronized		xxx$/T	
Kaolin	see clay				
Kyanite	raw concentrate			xx–xxx$/T	
	calcined			xxx$/T	
Lead	metal (bullion)			xx¢/lb	
Leucoxene			87% TiO_2	xxx$/T	max. 1% ZrO_2
Lime	chemical and industrial			xx$/T	
	construction grade			xx$/T	
	refractory dolomite			xx$/T	
	agricultural lime			xx$/T	
Lithium	petalite concentrate	−200 mesh	3.5/4.5% Li_2O	xxx$/T	
	spodumene concentrate		4/7% Li_2O	xxx$/T	
	lithium compounds			x$/lb	
	metal	ingot		xx$/lb	1000 lb lots
Magnesite	crude	lump		xxx$/T	
	agricultural grade	calcined powder		xxx$/T	
	industrial grade	calcined powder		xxx$/T	

Appendix III. Market Specifications

Commodity	Form Grade	Size	Basis	Price	Remarks
	maintenance grade	calcined powder		xxx$/T	
	brickmaking grade	calcined powder		xxx$/T	
Manganese	crude ore	bulk	48% Mn	x$/lb	long ton unit
	chemical grade	bulk	74/84% MnO_2	xxx$/T	
	battery grade	bulk	78/85% MnO_2	xxx$/T	
	ferromanganese		78% Mn	xxx$/T	
	metal			xx¢/lb	
Mercury	metal			xxx$/76 lb	flask
Mica	block, film, splittings			x$/lb	
	wet ground	powder		xxx$/T	
	dry ground	powder		xx$/T	
Molybdenum	ore concentrate		cont. Mo	x$/lb	
	technical grade oxide	powder	cont. Mo	x$/lb	
	ferromolybdenum	lump	cont. Mo	x$/lb	
Monazite			55/60% REO	xxx$/T	incl. ThO_2
Nepheline	glass grade	30 mesh		xx$/T	
syenite	ceramic grade	200/325 mesh		xx$/T	
Nickel	metal		cont. Ni	x$/lb	
	ferronickel			x$/lb	
	nickel oxide	sinter		x$/lb	
Niobium	see columbium				
Olivine	bulk, crushed	graded		xx$/T	
	foundry sand			xx$/T	
Perlite	raw, crushed	graded		xx$/T	
	expanded, milled			xxx$/T	
	aggregate, expanded			xxx$/T	
Phosphate	land pebble, bulk	unground	60/77 BPL	xx$/T	all prices negotiated
Platinum	platinum metal		cont. Pt	xxx$/oz	
Group metals	palladium metal		cont. Pd	xx$/oz	
	rhodium metal		cont. Rh	xxx$/oz	
	iridium metal		cont. Ir	xxx$/oz	
	ruthenium metal		cont. Ru	xx$/oz	
	osmium metal		cont. Os	xxx$/oz	
Potash	potassium chloride	bulk	60% K_2O	xxx$/T	
Pyrophyllite	refractory grade	bulk		xx$/T	
	ceramic grade	bulk		xx$/T	
	filler grade	bulk		xx$/T	
Quartz	foundry sand	bulk		xx$/T	
	glass sand	bulk		x$/T	

Commodity	Form Grade	Size	Basis	Price	Remarks
Radium	unencapsuslated			xx$/milligram	
Rhenium	metal	powder	cont. Re	xxxx$/lb	
	perrhenic acid			xxx$/lb	
Rutile	mineral concentrate	bulk	95% TiO$_2$	xxx$/T	
Salt	rocksalt	ground		xx$/T	10 ton lots
Sand	see quartz				
Scandium	metal		cont. Sc	xx$/gram	
	oxide		cont. Sc$_2$O$_3$	x$/gram	
Selenium	commercial grade	powder	99.5% Se	xx$/lb	
	high purity grade	powder	99.9% Se	xx$/lb	
Sillimanite	mineral concentrate	ground	70% Al$_2$O$_3$	xxx$/T	
Silver	metal		cont. Ag	xxxx¢/oz	
Slate	bulk powder	90/300 mesh		xx$/T	
Strontium	see celestite				
Sulfur	liquid			xxx$/T	
	solid			xxx$/T	
Talc	cosmetic grade	−325 mesh		xxx$/T	
	normal grade	200/325 mesh		xx$/T	
	ceramic grade			xx$/T	20 ton lots
Tantalum	tantalite concentrate		60% Ta$_2$O$_5$	xxx$/lb	
Tellurium	metal	powder	99/99.5% Te	xx$/lb	
	tellurium dioxide	powder	75% Te	xx$/lb	
Thallium	metal			x$/lb	
Thorium	oxide	powder	99.99% ThO$_2$	xx$/lb	
	nuclear grade ThO$_2$	powder		xx$/lb	
	metal	pellets	cont. Th	xx$/lb	
Tin	metal		cont. Sn	x$/lb	
Titanium	ilmenite concentrate		cont. TiO$_2$	xx$/T	
	rutile concentrate		cont. TiO$_2$	xxx$/T	
	titanium slag			xxx$/T	
	metal	sponge		x$/lb	
	TiO$_2$ pigment			xx¢/lb	
Tungsten	mineral concentrate			xxx$/T	
	metal		99.9% W	xx$/lb	
	ferrotungsten			xx$/lb	
Uranium	depleted uranium			x$/kg	sales controlled
	derby metal			x$/kg	sales controlled

Appendix III. Market Specifications

Commodity	Form Grade	Size	Basis	Price	Remarks
Vanadium	metallurgical grade oxide, fused		98% V_2O_5	x$/lb	
	technical grade oxide			x$/lb	
	ferrovanadium		cont. V	x$/lb	
Vermiculite	crude	bulk		xxx$/T	
Wolfram	see tungsten				
Wollastonite		300 mesh		xxx$/T	3 ton lots
Zinc	metal, various grades			xx¢/lb	
Zircon	foundry grade		65% ZrO_2	xxx$/T	
	premium grade		66% ZrO_2	xxx$/T	

Notes and abbreviations:
x, xx, xxx etc. = units, tens, hundreds.
Prices generally f.o.b. source. $ and ¢ U.S.
T not specified as to short, long, or metric ton.
oz = troy ounce.
cont. = contained.
REO = rare earth oxides.
BPL = bone phosphate lime 1% BPL = 0.458% P_2O_5.

Appendix IV

Units and Conversions

To convert	Into	Multiply by
	Length	
chain, surveyors (ch)	links	100
	rods	4
	feet	66
inches (in)	centimeters	2.54
	millimeters	25.4
kilometers (km)	feet	3281
	meters	1000
	miles	0.6214
meters (m)	centimeters	100
	feet	3.281
	kilometers	0.001
microns (μ)	meters	10^{-6}
miles	kilometers	1.609
	feet	5280
mils (mil)	inches	0.001
rods (rd)	chain, surveyors	0.25
	feet	16.5
	meters	5.029
	links	25
	Area	
acres (A)	square chains	100
	square feet	43560
	hectares	0.4047
	square meters	4046.9
hectares (ha)	acres	2.471
	square feet	1.075×10^5
	square meters	10^4
square kilometers (km^2)	square miles	0.3861
square meters (m^2)	square yards	1.1960
square miles (mi^2)	square kilometers	2.590
square yards (yd^2)	square meters	0.8361

Appendix IV. Units and Conversions

To convert	Into	Multiply by
	Volume	
cubic centimeters (cm³, cc)	cubic inches	0.610
	cubic meters	10^{-6}
	liters	0.001
	milliliters	0.9999
cubic feet (ft³)	cubic inches	1728
	cubic meters	0.0283
	liters	28.32
	quarts (US liquid)	29.92
cubic meters	cubic centimeters	10^6
	cubic feet	35.31
	cubic yards	1.3079
	gallons (US liquid)	264.2
	liters	1000
cubic yards (yd³)	cubic feet	27
	cubic meters	0.76455
gallons, US (gal)	cubic meters	3.785×10^{-3}
	liters	3.785
	gallons, imperial	0.83267
liters (l)	cubic centimers	1000
	cubic meters	0.001
	quarts (US liquid)	1.057
	Weight	
carats, metric (c)	milligrams	200
	grams	0.20
gallons of water	pounds of water	8.3453
grams (g, gm)	kilograms	0.001
	milligrams	1000
	ounces, troy	0.3215
kilograms (kg)	grams	1000
	long tons	9842×10^{-4}
	short tons	1.102×10^{-3}
micrograms (μg, γ)	grams	10^{-6}
milligrams (mg)	grams	0.001
ounces, avoirdupois (oz. av)	ounces, troy	0.9115
ounces, troy (oz. tr)	grams	31.1035
	ounces, avoirdupois	1.0971
	pennyweights, troy	20
	pounds, troy	0.0833
pennyweights, troy (dwt)	ounces, troy	0.05
	grams	1.5552
pounds, avoirdupois (lb)	grams	453.592
	pounds, troy	1.2153
	ounces, troy	14.5833
	short ton	0.0005
pounds, troy (lb. tr)	grams	373.242
	ounces, troy	12
	pennyweights, troy	240
	pounds, avoirdupois	0.8229

To convert	Into	Multiply by
tonnes (t) or metric ton	kilograms	1000
	pounds	2204.2
	tons, short	1.1023
tons, long (tn. l)	kilograms	1016
	pounds, avoirdupois	2240
	tons, short	1.120
	tonnes	1.016
tons, short (tn. s)	kilograms	907.185
	pounds, avoirdupois	2000
	tonnes	0.9072
Density		
grams per cubic centimeter (g/cm^3)	pounds per cubic foot	62.43
	pounds per cubic inch	0.03613
kilograms per cubic meter (kg/m^3)	grams per cubic centimeter	0.001
	pounds per cubic inch	3.613×10^{-5}
	pounds per cubic inch	0.0624
pounds per cubic foot (lb. av/ft^3)	pounds per cubic inch	5.787×10^{-4}
	grams per cubic centimeter	0.01602
	kilograms per cubic meter	16.018
pounds per cubic inch (lb. av/in^3)	grams per cubic centimeter	27.680
	kilograms per cubic meter	2.768×10^4
Concentration		
carat, karat—one part of gold to 24 of mixture	milligrams per gram	41.667
	parts per million	41666.7
	percent	4.1667
grams per tonne (g/t)	parts per million	1
	percent	0.001
	milligrams per kilogram	1
milligram per assay ton	troy ounces per short ton	1
milligrams per liter (mg/l)	parts per million	1
ounce, troy per ton, short (oz. tr/tn. s)	parts per million	34.286
parts per million (ppm)	grams per short ton	0.9072
	grams per tonne	1
	troy ounces per short ton	0.02917
	weight percent	10^{-4}
percent (%)	parts per million	10^4

Subject Index

Accuracy, 103
Acidity, 65
Acquisition value, 167
Ad valorem tax, 187
Aerial photography, 89, 92
Amalgamation, 142
Anion, 4
Annuity, 168
Aquaclude, 51
Aquifer, 51
Area measurement, 99
Arkansas, 36
Artesian flow, 55
Assay, 102
Assay wall, 100
Augering, 96
Aureole, 45, 100
Autogenous grinding, 136

Back arc basin, 23
Bacteria, 76
Bacterial stimulation, 131
Bacteriogenesis, 76, 77
Ball mill, 136
Banded iron formation, 41
Batholith, 35
Bayer process, 200
Beach placer, 203
Belt conveyor, 148
Beneficiation, 135
Besshi-type deposit, 69
Bias, 103
Bingham Canyon, UT, 77
Black sands, 203
Blasting, 118
Bonds, interatomic, 3, 4, 15
Bore hole extraction, 118, 128
Bouganville Copper Ltd., 149
Breccia pipe, 70
Broken Hill District, Australia, 47, 68
Buchans, Newfoundland, 77
Bureau of Land Management, 184
Bushveld Igneous Complex, South Africa, 31

Capillary fringe, 49
Capital formation, 165
Carajas, Brazil, 80
Carat, 206

Carlin, NV, 80
Cartel, 163
Cash flow, 165
Cation, 4
Central African Copperbelt, 42
Channelization, 55, 59, 61
Claim system, 182
Chattanooga Shale, 77
Chocaya, Bolivia, 71
Chuquimata, Chile, 68
Cincinnati Arch, 62
Classification, size, 13, 135, 137, 138
Cleavage, 9
Climax, CO, 122
Cobbing, 15, 140
Closed system, 16
Colorimetric testing, 88
Color of minerals, 15
Colquijirca, Peru, 71
COMEX, 164
Comminution, 10, 135
Commodity classification, 110
Complex ions, 65, 204
Complexing agents, 142
Component, 17
Concession system, 182
Cone crusher, 136
Coning and quartering, 102
Connate water, 49, 53, 56, 62
Consortium, 170, 195
Coppermine River, Canada, 42
Copper porphyry deposits, 60
Coproduct, 147
Cornwall District, England, 46, 161
Cost depletion, 187
Costeaning, 104
Cross sections, 100
Crushing, 10, 136
Crystal settling, 31
Crystal structure, 16
Cumulative curve, 14
CXHM test, 88
Cyaniding, 131, 206
Cyclone, 138
Cyprus-type deposit, 69

Darcy's law, 52
Dendritic growth, 17
Density, 14, 100

247

Depletion allowance, 187
Diamond drilling, 96
Dike, 36, 60
Discount rate, 167
Dredging, 123
Drilling, 96
Duty, 151

Economics, 161
  Annuity, 168
  Capital formation, 165
  Cash flow, 165
  Discount rate, 167
  Financial planning, 177
  Interest, 163, 166
  International viewpoint, 193
  Lifetime, 176
  Risk, 165, 166
  Risk assessment, 171, 172, 173
Eh, 75
Eh-pH diagram, 21
Electromagnetic radiation, 91
Elutriation, 12, 139
Epigenetic deposits, 21
Epithermal deposits, 69
Equipotential surface, 54
Erosion, 37, 87
Ertsberg Project, Indonesia, 149
Eutectic point, 19
Evaporite deposits, 39
Exploration funding, 165
Exploration geology, 80
Exploration tools, 80
Extraction methods, 118
  Bore hole extraction, 118, 128
  Dredging, 123
  Heap leaching, 1312
  Open pit mining, 118, 120
  Quarrying, 120
  Underground mining, 123

Fabric, 29
False color, 94
Ferromagnesian, 29
Fertilizer, 199
Field capacity, 49
Financial planning, 177
Fineness, 206
Finishing, 135, 148
Fissure filling, 70
Flin Flon, Canada, 77
Florida phosphate, 199
Flotation, 16, 141
Flow channel, 54
Flow line, 54
Flow sheet, mill, 135
Fluid inclusion, 76

Foreign operations, 195
Free energy, 16
Free-milling ore, 11, 137
Froth flotation, 16, 141

Gadsen Purchase, 183
General Mining Law of 1872, 185
Geochemical prospecting, 87
Geologic boundaries, 100
Geologic evaluation, 97
Geologic mapping, 81
Geophones, 84
Geophysical exploration methods, 81
  Electrical methods, 85
  Electromagnetic methods, 86
  Magnetic methods, 82, 206
  Seismic methods, 84
Geothermal fields, 53
Glory hole, 120
Gold prices, 206
Gold standard, 206
Goodnews Bay, AL, 202
Gossan, 73
Grade, 98, 101
Grain size, 11
Gravimeter, 85
Gray scale, 94
Grinding, 11, 136
Grinding efficiency, 151
Grizzly, 138, 153
Groundwater, 49
  Channelization, 55, 59, 61,
  Chemistry, 67
  Deep, 55
  Flow, 54
  Pore pressure, 59
  Shallow, 53
Guanajuato, Mexico, 70
Gulf Coast sulfur, 77
Gyratory crusher, 136

Habit, 16
Hardness, 9
Hazen Uniformity Coefficient, 14
Heap leaching, 131, 143, 206
Hemlo, Canada, 80
Histogram, 13, 88
Hockley Dome, LA, 77
Hot springs, 70
Huanchaca, Bolivia, 71
Humphrey spiral, 141
Hurdle rate, 168, 171
Hydraulic conductivity, 50
Hydrometallurgy, 15, 142
Hydrothermal fluid, 22, 34, 46, 65, 66
Hydrothermal ore deposits, 67, 204
  Classification, 68

Subject Index

Hyperfusible component, 29
Hypogene deposits, 21
Hypothermal deposits, 68

Igneous bodies, 35
Igneous rocks, 25, 28, 35
    Classification, 29
    Porosity, 50
Impregnation, 67
In-place value, 104, 161
Interest, 163, 166
International Monetary Fund, 206
Ionic bond, 4
Ionic structure, 4
Iron formation, 37
Island arc, 23
Isometric block diagrams, 100

Jaw crusher, 136
Jigging, 138
Joint Venture, 170
Juvenile water, 49, 56, 58

Kidd Creek, Canada, 77, 80
Kimberlite, 35
Kolm, Sweden, 77
Kupferschiefer, Europe, 42, 77
Kuroko-type deposits, 69, 77

Landsat, 1, 93
Land use in the U.S., 184
Lattice, 16
Lava, 28
Lava plateau, 36
Lifetime, 117, 150, 161, 176
Lithification, 39
Lithosphere, 22
LIX, 143
Lognormal distribution, 88
London Metal Exchange, 164
Louisiana Purchase, 183
Luster, 15

Magma, 18, 22, 33, 45, 60, 69
    Crystallization, 28, 56
Magnetometer, 83
Manner of breaking, 9
Marketing, 145, 194
Marshall Plan, 206
Merensky Reef, South Africa, 32
Mesothermal deposits, 68
Metacryst, 45
Metal prices, 174
Metamorphic rocks, 25, 44
Metamorphism, 22, 44
    Contact, 46
    Metamorphic facies, 44
    Metamorphic grade, 44
    Regional, 46
    Water released by, 58
Meteoric water, 49
Mill circuits, 135, 150
    Head to tail design, 150
    Tail to head design, 151
Milling, 28, 117, 135
    Circuits, 135, 150
    Flow sheet, 135
    Operations, 135
    Principles, 135, 150
    Processes and devices, 135
    Solvent exchange, 143
    Waste disposal, 154
Milling operations, 135
    Beneficiation, 135
    Classification, 135, 138
    Comminution, 135
    Finishing, 135, 148
    Separation, 135, 141, 144
    Transportation, 148
Mine openings, 118, 124, 125
Mine planning, 118
Mineral, 3
    Abundance, 3
    Association, 21
    Classification, 4
    Composition calculations, 9
    Defined, 3
    Habit, 16
    Interaction with light, 7
    Magnetic and electrical properties, 15
    Physical properties, 5, 9
Mineral deposits, 22, 28
    Classification, 106
Mineral exploration, 79
Mineral law, 182
Mineral Leasing Act of 1920, 185
Mineralogy, 3
Mine support, 123, 124
Minimum Anticipated Cumulative Demand, 192
Mining, 28, 117
    Law, 185
    Organizations, 170
    Regulations, 186
Mining and Minerals Policy Act of 1970, 185
Mining, underground. See underground mining
Mining width
Mississippi Valley-type deposit, 71
Monopoly, 163
Mount Isa, Australia, 77, 125

Nairne, Australia, 77
National Park Service, 184

Obduction, 23
Olympic Dam, Australia, 80
Open pit mining, 120
Open system, 16
Ore, 3, 97, 191
   Deposition, 74
   Production, 192
   Solution and deposition, 62, 76
Orebody, 25, 104, 118
Ore deposit, 3, 24, 28
Ore guide, 81
Oriskany Sandstone, WVA-PA, 40
Overburden, 120
Overburden ratio, 120
Ownership, 182
Oxidation-reduction, 20, 75

Panning, 87, 206
Paragenetic deposits, 21
Paramarginal deposit, 191
Partition, 29
Pathfinder elements, 88, 206
Payback period, 171
Pegmatite, 32, 45, 204
Percentage depletion, 187
Permeability, 51, 61
   Differential, 55
pH, 21, 75
Phase diagrams, 17
Phase Rule, 17
   Of Gibbs, 17
   Of Goldschmidt, 17
Phosphoria Formation, 43, 199
Phosphorite, 42, 185, 199
Phreatic zone, 49, 67, 72
Pinto Valley mill, AZ, 149
Pipe, 35
Pixel, 94
Placer deposit, 24, 38, 68, 123, 203, 205
Place value, 150, 198
Plate tectonics, 22
Plutonic rock, 35
Pneumatolitic minerals, 34
Polymorph, 17
Porosity, 50
Porphyry copper deposits, 69
Potentiometric surface
Potosi, Bolivia, 71
Precision, 103
Prediction, 171, 172
Pregnant pond, 132
Pressure, 59, 76
   Hydrostatic, 60
   Lithostatic, 60
Production tax, 187
Profitability, 97, 170, 172
Property tax, 187
Prospecting methods, 81
   Geochemical, 87
   Geologic mapping, 81
   Geophysical, 81
   Remote sensing, 89
P-T diagrams, 17
Public lands, 183, 197
Pyroclastic, 36

Quarrying, 120

Radical, 4
Rake classifier, 138
Reaction series, 29
Reclamation, 154
Red bed-copper association, 42, 77
Redox, 20, 75
Red Sea, 24
Regulations, 154, 186
Remote sensing, 61, 88, 89, 206
   Aerial photography, 92
   Satellite imagery, 93
   Side-looking radar, 94
   Stereoscopic vision, 92
Reserve, 191
Resistivity, 85
Resistivity drilling, 85
Resource, 191
Rest magma, 29, 33
Rifting, 24
Riparian laws, 182
Risk, 165, 166, 172
Risk assessment, 171, 172, 173
Rock, 3
   Igneous, 25, 28, 35
   Metamorphic, 25
   Sedimentary, 25, 39, 74
Rock cycle, 24
Rock failure, 59, 60
Rod mill, 136
Royalty, 183, 187
Ruth, NV, 122

Saltation, 37
Salting, 101
Salt domes, 76
Samarco Mineracao, Brazil, 149
Sampling, 101, 135
Satellite imagery, 93
Savage River Mines, Tasmania, 149

Subject Index

Screening, 11
Screens, 137, 138
Second boiling, 56
Sedimentary rocks, 25, 39, 74
Seismic prospecting, 84
Seismic pumping, 60
Selective melting, 22
Separation, 135, 144
    Devices, 141
    Principles, 144
Severance tax, 187
Shaking table, 141
Shielding, 3
Shipping cost, 198
Side-looking radar, 95
Sierra Nevada, CA, 38
Sieves, 12
Sill, 36, 60
Siltation, 155
Silt ponds, 155
Sinking fund, 170
SIX, 143
Skarn, 45
Skellefte, Sweden, 77
Slime pond, 155
Slurry pipeline, 148
Solid solution, 5
Solubility, 15
Solvent exchange, 143
Sorting, 12
Specific gravity, 14
Splitting, 101
SPOT satellite, 93
Spreading center, 22
Standard potential, 20
Statistical variability, 103
Stereoscopic vision, 92
Stillwater Complex, MT, 202
Stockpile, 196, 201
Stockwork, 67, 70
St. Peter Sandstone, Il-MO, 40
Strategic metal, 201
Streak, 15
Stripping ratio, 121
Subduction, 22
Sublimation, 35
Submarginal deposit, 191
Sullivan, Canada, 77
Supergene deposits, 21
Supply and demand, 162
Surge storage, 153
Suspensoid, 14
Syngenetic deposits, 21

Tactite, 45
Tapira, Brazil, 149
Targeting, 79
Taxation, 187
    Impact, 189
    Income, 187, 189
    Production, 187, 189
    Property, 187, 189
Telethermal deposits, 71
Temperature determination, 76
Tenor, 98
Tonnage calculation, 100, 104, 145
Transportation, 148, 198
Treaty of Guadalupe Hidalgo, 183
Trenching, 104
Triple Point, 17
Trombetas, Brazil, 80
T-X diagrams, 18

Underground mining, 123
    In advance, 126
    In retreat, 126
    Mine openings, 124
    Mine support, 124
    Principles, 123
Underground mining methods, 123
    Block caving, 126
    Cut and fill stoping, 126
    Room and pillar, 125
    Shrinkage stoping, 126
    Sublevel caving, 126
Unit depletion, 187
Upper Peninsula, MI, 46
Uranium-vanadium deposits, 71
U.S. Forest Service, 184
U.S. Mining law, 184

Vein, 36, 45, 66, 70, 97, 100
VLF method, 86
Void ratio, 50
Volcanism, 34, 69
Volume, 98

Waste disposal, 148, 154
Wasting asset, 161, 169
Water, 33, 56
    Connate, 49, 53, 56, 62
    In magma, 33, 56
    Juvenile, 49, 56, 58
    Meteoric, 49
    Sills and dikes, 60
    Underground, 49
Water rights, 182
Water table, 49, 54, 67, 73, 74
    Chemical reactions at, 67
Weathering, 36, 39, 72, 87
Weipa, Australia, 80

Wettability, 16
White Pine District, MI, 42, 77
Wilderness Act, 184
Wilfley table, 141
Winfield Dome, LA, 77
Witwatersrand, South Africa, 68, 204

Xanthate, 16, 142

Yakutsk Republic, U.S.S.R., 36
Ytterby, Sweden, 33

Zambian Copperbelt, 77

Mineral and Commodity Index

Acanthite, 70
Adularia, 70
Alumina, 200
Aluminum, 200
Alunite, 70
Amphibole, 29, 33
Andalusite, 17
Anhydrite, 76, 85
Anorthite, 19
Antimony, 68, 195
Apatite, 43
Arsenic, 46
Arsenopyrite, 67
Asbestos, 46

Baddeleyite, 203
Barite, 62, 66, 75
Bauxite, 37, 80, 140, 163, 200
Beryl, 15, 33
Beryllium, 33
Biotite, 29, 45, 203
Bismuth, 46
Bismuthinite, 67
Borates, 39

Cadmium, 68
Calcite, 4, 15, 68, 75
Cassiterite, 38, 67
Cesium, 33
Chalcocite, 73
Chalcopyrite, 8, 67, 72
Chlorite, 45, 68, 70
Chromite, 15, 23, 32
Chromium, 32, 37
Cinnabar, 70, 141
Clay, 37, 39, 50, 68, 185, 198, 200
Coal, 39
Cobalt, 24, 46, 68, 163, 195
Cobalt-nickel arsenides, 67
Columbium, 33, 38
Copper, 21, 23, 32, 46, 68, 69–72, 77, 80, 131, 143, 195
Corundum, 5, 8, 46
Cuprite, 21

Dawsonite, 200
Diamond, 4, 15, 35, 38, 87, 143, 163, 164, 195
Diopside, 19

Dolomite, 68
Dolostone, 39

Electrum, 70, 205

Feldspar, 33, 45
Ferromagnesian, 29
Fluorine, 199
Fluorite, 15, 71

Galena, 5, 62, 67
Garnet, 33, 45, 46
Glass sand, 40, 98
Gold, 15, 23, 24, 38, 67, 68, 70, 79, 80, 87, 88, 123, 131, 142, 143, 204
Graphite, 39, 46, 195
Gypsum, 39, 85, 98

Halite, 4, 62
Hematite, 5, 37

Ilmenite, 203
Iridium, 202
Iron, 40, 46, 195
Iron ore, 31, 412, 80, 163, 172, 195
Iron oxide, 37, 72, 77, 140

Jadeite, 24

Kyanite, 17, 45, 46

Lead, 23, 24, 46, 68, 164
Lepidolite, 203
Limestone, 39, 98
Lithium, 33
Loellingite, 67

Magnesium, 164
Magnetite, 15, 31, 32, 83
Manganese, 23, 24, 37, 46, 77
Mica, 15, 33, 45, 203
Molybdenite, 8
Molybdenum, 8, 23
Monazite, 203
Muscovite, 29, 203

Nepheline, 29
Nepheline syenite, 200
Nickel, 23, 24, 32, 37, 46

253

Olivine, 8, 29
Orthoclase, 68
Osmium, 202

Palladium, 202
Pentlandite, 67
Phlogopite, 203
Phosphate, 40, 43, 185, 195, 199
Plagioclase feldspar, 29
Platinum, 22, 23, 38, 202
Platinum group metals, 202
Potash, 185
Potash feldspar, 29
Pyrite, 67, 68, 70
Pyrophyllite, 46
Pyroxene, 29
Pyrrhotite, 15, 67

Quartz, 29, 33, 37, 40, 45, 68, 71

Rare earths, 33
Rhenium, 8
Rhodium, 202
Roscoelite, 203
Ruby, 8
Ruthenium, 202
Rutile, 203

Salt, 39, 131, 185
Sand and gravel, 185, 198
Scheelite, 15, 67, 200
Sericite, 68, 70, 203
Sillimanite, 17, 45

Silver, 15, 23, 24, 46, 68, 70, 80, 142
Sphalerite, 62, 76, 75
Staurolite, 45
Stibnite, 70
Stone, 24, 185
Sulfur, 76, 96, 131, 185

Tantalum, 33, 38
Thorium, 33, 164
Tin, 23, 24, 33, 38, 46, 80, 87, 123, 163, 164, 195
Titanium, 164
Topaz, 33
Tourmaline, 33
Tungsten, 23, 24, 33, 46, 195, 201

Uranite, 67, 75
Uranium, 15, 24, 33, 46, 71, 77, 80, 131, 143, 164, 199

Vanadium, 71
Vermiculite, 203

Wolframite, 67, 200

Xenotime, 203

Yttrium, 33

Zeolite, 70
Zinc, 23, 24, 68
Zircon, 203
Zirconium, 164

About the Author

Dr. William H. Dennen is professor emeritus, Department of Geology, University of Kentucky. His academic teaching and research career has spanned four decades at the Massachusetts Institute of Technology and Kentucky and has included mineral exploration and evaluation activities on five continents.

Son of a mining engineer, he received his bachelor's degree in economic geology, minor in mining, from the Massachusetts Institute of Technology in 1942 and served for the ensuing four years as an anticraft artillery officer in the U.S. Marine Corps. His PhD (petrology) was conferred by MIT in 1949 and he served on the geology faculty there until 1967 when he became Professor and Chairman of the Department of Geology at the University of Kentucky.

The author is an accomplished spectrochemist and served as Director of the Cabot Spectrographic Laboratory from 1954 until his retirement from active teaching in 1985. He has done geochemical exploration and bedrock mapping for the U.S. Geological Survey and a number of private, corporate, and governmental bodies, published 70 technical papers and maps, and is the coauthor of *Geology and Engineering* and *Principles of Mineralogy*. Work in the field has required hands-on application of many of the techniques and procedures discussed in the text.

Professor Dennen has served as director for two small corporations, is a Fellow of the Geological Society of America, member emeritus of the Society for Applied Spectroscopy, and a Certified Professional Geological Scientist (#6533). He currently resides with his wife in Nahant, Massachusetts and is associated with Salem State College as a Visiting Scientist.